Infrared
Spectroscopy
of
High Polymers

Rudolf Zbinden

Pioneering Research Division
Textile Fibers Department
E. I. duPont de Nemours & Co., Inc.
Wilmington, Delaware

1964

ACADEMIC PRESS · New York and London

ACADEMIC PRESS INC.
111 Fifth Avenue, New York 3, New York

United Kingdom Edition published by
ACADEMIC PRESS INC. (LONDON) LTD.
Berkeley Square House, London W.1

LIBRARY OF CONGRESS CATALOG CARD NUMBER: 64-14570

PRINTED IN THE UNITED STATES OF AMERICA

**Infrared
Spectroscopy
of
High Polymers**

Preface

The purpose of this book is to assist both the spectroscopist in interpreting spectra of high polymers and the polymer scientist in using infrared spectroscopy for the elucidation of chemical and physical structure of polymers. The book contains a discussion of methods and fundamentals, but is not a catalog of spectra.

Chapter I is a general survey of the subject while the remaining four chapters are more thorough treatments of some aspects of spectroscopy of high polymers. A guide fo the literature and to sources of many spectra is contained in the appendix.

It is a pleasure to acknowledge the support of various members of the Textile Fibers Department of E. I. duPont de Nemours & Company, Inc. I would like to thank especially Dr. G. F. Lanzl who suggested the writing of this book and Dr. J. R. McCartney for continual interest and encouragement. I am grateful to Drs. L. F. Beste and C. R. Bohn for reading parts of the manuscript and making suggestions for its improvement, and to Dr. D. F. Akeley for writing computer programs and carrying out several numerical computations. Many thanks are due to Miss T. A. Balisco, Mrs. M. C. Evlom, Miss J. M. Iannarone, Miss M. J. Thompson, and Mrs. H. H. Smith for tedious typing work and to Mr. J. J. Duffy, Mr. F. A. Schuck, and Mr. N. F. Van Hove for preparing all the figures. I also wish to express my appreciation to the Management of the duPont Company for permission to write this book.

I am indebted to my wife for patience and moral support, for reading the manuscript as well as the proofs, and for many suggestions improving my English.

March, 1964   **Rudolf Zbinden**[*]

[*]Present address: J. R. Geigy A.G., Basel, Switzerland.

Contents

III

Numerical Calculations of Vibrations in Chain Molecules

IV

Vibrational Interaction in Chain Molecules

V

Orientation Measurements

Appendix

—I—

Characteristic Features of Polymer Spectra

1. The Structure of Polymers

The purpose of this first section is to describe briefly the chemical and physical structure of high polymers. Such a description, in a few pages, cannot be a comprehensive one. It will merely serve as an introduction to the main subject of this book. A large number of textbooks and several handbooks are available for much more detailed information (see e.g. Refs. 6, 34, 41, 52, 75, 78).

A polymer consists of macromolecules, which are composed of a multiple of relatively small repeat units. In cellulose, for instance, the repeat unit is $C_6H_{10}O_5$, and one molecule might contain several thousand such units. The molecular weight of polymers is usually in the range 10^3–10^6. It usually is not possible to give a complete chemical description of a polymer since any proposed simple structure of macromolecules can be only approximately correct. Deviations from such idealized structures can occur in many ways. In a compound such as benzene or methane, all the molecules are exactly alike. For a polymer this is usually not so. A given sample contains a wide variety of molecules of different molecular weight. Furthermore a molecule is not always an exact multiple of the repeat unit. Polypropylene, for instance, is not exactly $-(-CH_2-CH-)-_n$, where n is an integer, with CH_3 since it contains end groups A and B so that the actual structure has the form $A-(-CH_2-CH-)_n-B$, with CH_3. The nature of the end groups in this case depends on the method of polymerization. Most of the so-called linear polymers are not perfectly linear—in reality they are branched. A typical example is polyethylene. The early experimental and commercial products consisted of highly branched molecules. Only with very special polymerization techniques was it possible to prepare chains that were almost linear. Chains may also have other irregularities. If vinylidene chloride, $CH_2=CCl_2$, is polymerized, the monomer units can add in a "head-to-head" or in a "head-to-tail" fashion, and in practice both species are usually present in one molecule. The relative ratio of the two depends on the polymerization conditions. These conditions can also influence the structure in an even more refined way. Depending on the catalyst, a "tactic" or an "atactic" chain can be formed. Let us consider the head-to-tail polymer of styrene, $C_6H_5CH=CH_2$, as an example. Each repeat unit contains an asymmetric carbon atom and the substituents of this carbon atom may be arranged in two different configurations that

1

are mirror images of each other. They correspond to the *d*-form and the *l*-form, respectively, of optically active compounds. If the two forms are distributed at random along the chain, the polymer is atactic, but if a chain consists of either all *d*-units or all *l*-units, the polymer is isotactic. Other types of tacticities are possible for a large number of polymers but they will not be discussed at this point. A review of this subject has been given by Natta (Ref. 58).

The geometrical shape of a polymer molecule is another characteristic property. A linear polymer molecule in a dilute solution is not a straight chain. It rather assumes the form of a random coil, the shape of which is somewhat determined by the chemical structure of the chain. In the amorphous solid state the shape of the individual molecules is probably very similar to the random coil of a dilute solution, but this is different for a crystalline sample. The crystalline state in a polymer differs appreciably from a crystal of a nonpolymeric compound where virtually a hundred percent of the material can crystallize. In a "crystalline" polymer we find crystalline and amorphous regions. The crystalline regions (of the order 100 Å in size) are not perfect crystals but just areas in a highly ordered state with quite diffuse boundaries. An individual polymer chain may pass through several crystalline and amorphous regions. Recently it was discovered that some polymers are capable of forming "true" single crystals (Refs. 33, 48, 83). Polyethylene, precipitated from a dilute solution, can form diamond shaped platelets about 50 to 100 Å thick and 1 to 2 μ in diameter. The molecular chain axis is perpendicular to the plane of the platelet. At the platelet surface the chain is folded over on itself since the chains are much longer than the single crystal is thick. The shape of the chains in a crystal may vary for various polymers. In crystalline polyethylene, for example, the carbon skeleton is a planar zig-zag chain. Other polymers prefer to crystallize in the form of a helix. Isotactic polystyrene and polypropylene are examples (Ref. 58).

All the structural features discussed so far determine the physical properties of a polymer sample to quite a large extent. It is not our intention to give a comprehensive structure property relation picture, but we will just mention at random a few such correlations: The solution viscosity and melt viscosity depend on such variables as chemical structure, chain flexibility and, in particular, the molecular weight. The solubility of a polymer is highly determined by specific interactions of polymer and solvent, but the solubility may also depend upon the geometrical structure of a chain. For example, atactic polystyrene is very soluble in benzene, while the isotactic form is practically insoluble in the same solvent. The mechanical properties of a solid polymer sample depend particularly on the geometrical arrangement of the chain molecules within the sample. Under the proper conditions it is possible to stretch some polymers with the result of permanently aligning the molecular chains more or less parallel to the stretching direction. The tensile strength of such a sample measured in the direction of stretching may be as much as ten times as high as for the unoriented sample. This factor would be smaller for a lower degree of orientation. An almost linear polyethylene can be stretched to a higher orientation and therefore to a higher tensile strength than a sample containing highly branched molecules.

There are, of course, a large number of more subtle structure-property correlations involving structural parameters such as amount of crystallinity, crystallite size, void content, etc., not discussed here. Many correlations of this kind can be found in the literature (see, e.g., Refs. 15, 16, 17, 52).

2. Comparison of Spectra of Nonpolymeric and Polymeric Compounds

In the previous section we have briefly described some structural characteristics of polymers. From that description the differences between polymeric and nonpolymeric compounds are quite apparent. Some of these differences are reflected in the infrared spectrum. Our task in this book is to point out the characteristic features of polymer spectra, analyze them, and show how they can be useful in polymer science. We will discuss the kind of spectrum one would expect from a "perfect" or idealized polymer structure such as an infinitely long, straight chain composed of an infinite number of regularly arranged repeat units. We shall then describe the observed spectra and show how they deviate from the expected spectra of idealized models.

A. NONPOLYMERIC COMPOUNDS

An N-atomic molecule has $3N$ normal vibrations (including rotations and translations of the molecule as a whole). For relatively small and highly symmetrical molecules it is usually possible to correlate fully the observed vibrational infrared and Raman spectra with the expected normal vibrations (Ref. 36), while for a large molecule this task is much more difficult if not impossible, but despite this fact one can get useful information from an observed spectrum. It is possible to analyze partially such a spectrum since some normal vibrations involve quite localized motions within a molecule. They are the so-called "characteristic group vibrations." Vibrations that are predominantly carbon-hydrogen stretching modes, for example, fall in the spectral range of about 2800 to 3100 cm^{-1}, while carbon-oxygen stretching vibrations for carbonyl groups usually fall between 1640 and 1780 cm^{-1}. Analytical chemists are making extensive use of this fact. For them infrared spectroscopy has become a tool in the analysis of the chemical structure of molecules (Ref. 4).

The infrared absorption spectrum depends strongly on intermolecular interactions; therefore, the spectra of a compound in the gaseous and liquid state show pronounced differences. In the gaseous state the molecules are essentially free and the observed spectrum consists mainly of rotation-vibrational bands. Such spectra have been observed particularly for relatively small molecules where it is possible to analyze the rotational fine structure of a vibrational band. From this analysis structural parameters such as bond angles and interatomic distances can be obtained. The spectrum of a liquid usually does not contain any rotational fine structure because of molecular collisions that make a free rotation impossible. Irregular intermolecular interactions in a liquid cause shifts and broadening of vibrational absorption bands, but in the crystalline state the interactions are well defined and the same for each molecule. We therefore find a sharpening of bands by going from the liquid to the solid state. Only a relatively small number of infrared and Raman spectra of molecular and ionic crystals have been analyzed (see, e.g., Refs. 20, 21, 22, 37, 38, 39, 49, 54, 63, 64, 79, 80, 81, 87).

B. POLYMERIC COMPOUNDS

For the giant molecules of a polymer the number of atoms per molecule is very large, which leads to an enormous number of normal vibrations. It is, therefore, quite surprising that for most polymers the actual absorption spectrum is relatively simple (Ref. 51). Often it is not even possible to tell whether a spectrum of an unknown compound is that of a polymer or a small molecule. The reason for the relative simplicity of most polymer spectra is twofold. First it turns out that many of the normal vibrations have almost the same frequency and, therefore, appear in the spectrum as one absorption band only, and second the selection rules for polymers can be quite strict so that only a few of the many normal vibrations are infrared- or Raman-active. These points will now be discussed.

Consider as an idealized example a straight infinite chain of polystyrene where all the repeat units have exactly the same geometrical configuration.

repeat unit
$n = 16$

For a moment we make the hypothetical assumption that the repeat units are free, that is, all those carbon-carbon bonds are broken which are crossed by a dashed line. The repeat unit can then be considered as a 16-atomic "molecule" with $3n - 6 = 42$ normal vibrations. It has no symmetry; therefore, all the vibrations are infrared- and Raman-active. Even though this molecule is relatively small it would be difficult to carry out a normal coordinate analysis, predict all the vibrational frequencies, and assign them to observed absorption bands. Nevertheless it should be possible to identify a number of absorption bands with characteristic group vibrations for the $>CH_2$ group as well as for the monosubstituted benzene ring. We would expect, for example, to observe aliphatic and aromatic C—H stretching vibrations in the 2800–3000 cm^{-1} range and in the 3000–310^{-1}0 cm range, respectively. Let us now consider the actual polymeric chain where the broken carbon-carbon bonds are restored. Each repeat unit has still the same normal vibrations as before but they now are somewhat affected by the interaction with the tightly coupled adjacent repeat units. The effect of this interaction on the characteristic group vibrations is quite small, while the frequency of the vibrations that are predominantly motions of the aliphatic carbon atoms is affected appreciably. This problem of vibrational interaction in chain molecules will be discussed at length in Chapters III and IV. At this point we describe briefly how this interaction is reflected in the spectrum.

First we consider just two identical repeat units. If the coupling is extremely small or nonexistent each normal vibration is doubly degenerate, but larger interaction will cause each vibration to split into two, similar to the case of two coupled oscillators described in every textbook on elementary mechanics. If there are N

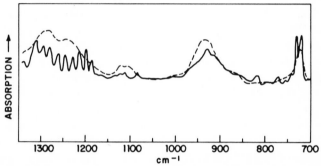

FIG. 1. Absorption spectra of heneicosanoic acid, $CH_3(CH_2)_{19}COOH$. After Sinclair, McKay, and Jones (Ref. 74). KEY: ——: crystalline state, as a nujol mull; ---: amorphous state, as a solution in CS_2, 0.046 M conc., cell thickness 1 mm.

coupled repeat units in the chain, each frequency will be split N-fold. The amount of splitting depends upon the degree of interaction. It is usually small for characteristic group vibrations but large for chain skeletal vibrations as discussed in the previous paragraph. For the characteristic vibrations of the benzene ring in polystyrene the splitting is too small to be observed, while the interaction of CH_2-group vibrations in normal hydrocarbon chains can cause quite appreciable splittings as shown in detail in Chapter IV. There we will find that individual absorption peaks of such a split band cannot be observed for polymers but only for compounds of a definite molecular weight and only when these compounds are in the crystalline state as illustrated by Fig. 1 for heneicosanoic acid (Ref. 74). The splitting of the band in the 1180–1310 cm^{-1} region is particularly pronounced (see also Refs. 61, 71). For a given polymer sample the splitting of an absorption band is different for chains of different length. The observed spectrum is a smeared out average of the individual peaks and, therefore, appears to be quite simple showing relatively little fine structure.

Another reason for the simplicity of polymer spectra is the existence of strict selection rules, particularly for idealized chains that are perfectly straight and infinitely long. Such chains can be considered as one-dimensional infinite crystals having an infinite number of normal vibrations. It will be shown in Chapter II that for such a system only $3n$ vibrations (including rotations and translations of the chain as a whole) are potentially infrared- or Raman-active (n is the number of atoms per repeat unit). Bhagavantam and Venkatarayudu (Ref. 5) were the first to give selection rules for molecular crystals. They showed that all the potentially active frequencies are obtained by examining a single unit cell, which in our case is a repeat unit. They specifically postulated that only those vibrations of a crystal would have to be considered for which corresponding atoms in all the unit cells move in phase. These are the so-called unit cell or factor group vibrations discussed in Chapter II. All the other vibrations are inactive as fundamentals in the infrared spectrum and the Raman effect. Depending on the symmetry properties of the crystal (or the chain) only part of the $3n$ unit cell vibrations may show some spectral activity.

The selection rules of crystals and chains can also be described in view of the vibrational interaction discussed in a previous paragraph. Consider a characteristic group vibration of a repeat unit. Due to coupling, this vibration will split into an infinite number of vibrations for an infinitely long chain. All the vibrations will fall in a frequency interval (the width of which increases with increasing coupling forces) but only one of the infinite number of vibrations is potentially infrared- or Raman-active. For chains of a finite length (N repeat units) a splitting into N frequencies is expected and it turns out that the unit cell vibration (all unit cells vibrate in phase) is by far the strongest one. All the others are either of zero intensity or relatively weak. This problem is treated quantitatively in Chapter IV.

From the discussion in Section 1 it is clear that actual polymer samples do not consist of infinitely long, straight chains with regularly arranged repeat units, and deviations from such an idealized model are reflected in the spectrum, where broadening as well as shifts of absorption bands may be observed. Intermolecular interaction can also affect the spectrum. In the amorphous regions this interaction between adjacent chain segments is random and a broadening of absorption bands is expected, but in crystalline areas the interchain coupling is regular and well defined; effects such as band splittings are expected. Such effects of crystallinity on the spectrum of a polymer are described in Section 4. Other characteristic features of polymer spectra will be apparent from the following discussion.

3. Orientation

A. General Principles of Infrared Dichroism

Consider an individual nonpolymeric molecule with known normal vibrations. For an infrared-active vibration the transition moment has a certain direction with respect to the geometry of the molecule. In a classical model this is the direction of the electric dipole moment change which is coupled with the vibration. In a hypothetical absorption experiment we arbitrarily fix the molecule in space with the transition moment in the z-direction. Absorption occurs only if the electric vector of the radiation has a component in that direction. Radiation with the electric vector perpendicular to the z-axis is not absorbed. The absorption intensity, therefore, depends on the angle of incidence and the direction of polarization. Unpolarized radiation in the xy-plane will be strongly absorbed while a beam along the z-axis is not absorbed at all, since for the latter the electric vector is perpendicular to the transition moment of the molecule. For polarized light the orientation effects are more pronounced. A beam in the xy-plane, for example, polarized with the plane of polarization perpendicular to the xy-plane, is absorbed very strongly since the electric vector is parallel to the transition moment, but no absorption is expected for a beam for which the xy-plane is the plane of polarization.

The molecules in a gas or a liquid are oriented at random in space. In an absorption experiment with such a sample no direction is distinguished from any other one and the absorption intensity, being the average of the absorptions of all the molecules, is independent of the angle of incidence or the direction of polarization of the radiation. For crystalline solids the situation is quite different. Let us consider a single crystal where all the molecules are oriented alike, that is, a crystal with a unit cell containing one molecule only. An absorption experiment with such a crystal is equivalent to the hypothetical case described in the previous paragraph, provided we can neglect the intermolecular interaction. Strong variations in absorption intensity are expected for different directions of incidence and, in particular, for different directions of polarization. A measurement for such differences is the "dichroic ratio," which is usually defined as the ratio of the optical densities of an absorption band measured with radiation polarized respectively parallel and perpendicular to a given direction in the crystal. More subtle definitions will be given below and especially in Chapter V.

To observe spectra of single crystals it is necessary that the crystals have a shape suitable for spectral analysis in the infrared. A platelet of 5 to 20 μ in thickness is usually required and, unless an infrared microscope is used, the diameter of the platelet should be a few millimeters to cover the light beam in a spectrometer. It is often quite difficult to obtain single crystals in this shape so that the number of nonpolymeric crystalline compounds analyzed by polarized light is rather limited (see, e.g., Refs. 20, 65, 86, 87).

Many spectra of crystalline compounds have been observed but mostly of samples in form of a nujol mull or a KBr pellet, techniques that do not lend themselves to orientation studies since the sample is ground very finely and the individual crystals are distributed randomly in the embedding medium. The situation is quite different for polymers, and it will be discussed in the following paragraph.

B. DICHROISM IN POLYMERS

Most of the linear polymers can be oriented permanently by some mechanical treatment such as stretching or shearing. This behavior is quite unique for polymers. A film strip of polyethylene can, for example, be stretched several-fold at about 120° C with the result that most polymer chains are oriented within an angle of less than 10° to the stretching direction. It is not possible to produce such a high degree of orientation for all polymers, but in general some orientation can be produced. It is important to note that in a polymer the orientation is never as perfect as in a single crystal. To characterize an imperfect orientation a distribution function is required, that is, a function that expresses the fraction of chain segments parallel to any direction in space. Such distribution functions are discussed in Chapter V.

It is often quite easy to prepare thin partially oriented films of polymers suitable for infrared analysis and polarization measurements as opposed to preparing a thin platelet of a single crystal from a nonpolymeric compound. This explains the relatively large number of polarization studies on polymers in the literature.

Since polymers in general are oriented with the chain axis preferentially parallel to the stretching direction, the geometrical configuration of the chains within a sample is approximately known. This is not so for single crystals of small molecules, where no *a priori* information is available about the orientation of the molecule with respect to a given face of the crystal. We now consider a sample of a linear polymer (a thin film or a fiber) stretched in one direction, where the stretching process produced partial axial orientation. Such a sample is mounted in a spectrograph with the stretching direction parallel to the slit. The absorption spectrum is then recorded for polarized light with the electric vector parallel and perpendicular to the stretching direction. For any particular band we can determine the absorption intensities (optical densities) $I_{\parallel}$ and $I_{\perp}$ for the two polarization directions, respectively. The ratio R of these two intensities

$$R = \frac{I_{\perp}}{I_{\parallel}}$$

is called the dichroic ratio for the particular absorption band. This definition is somewhat arbitrary and the inverse ratio is sometimes called the dichroic ratio. The value of R can range from zero (no absorption in the perpendicular direction) to infinity (no absorption in the parallel direction). If R is smaller than 1.0 the band is called a parallel band; if R is larger than 1.0 it is called a perpendicular band. This nomenclature refers not only to the stretching direction but also to the chain axis since the chains are preferentially oriented in the stretching direction. In most practical cases the observed dichroic ratios are between 0.1 and 10.

We now will discuss the two principle parameters that determine the dichroic ratio of an absorption band. One of them is the degree of orientation, the other one is the angle θ between the direction of the transition moment and the chain axis.

At this point we limit the discussion to a few specific examples and reserve a full treatment of the subject for Chapter V.

In the first example we consider the effect of orientation on the infrared dichroism for a vibration with a transition moment perpendicular to the chain axis ($\theta = 90°$). We would expect a dichroic ratio of infinity for a perfectly oriented sample and no dichroism ($R = 1$) for an unoriented sample. Depending on the degree of orientation, any value of R between one and infinity is possible. For the hypothetical case, for example where all chains form an angle of 20° with the stretching direction ($\gamma_0 = 20°$ in Fig. 20, Chapter V), we would expect a dichroic ratio of 8.1 [see Eqs. (39) and (50), Chapter V].

The influence of θ on the dichroic ratio is illustrated by the following example. We consider a sample with perfect axial orientation but where the transition moment can form an angle θ smaller than 90° with the chain axis. The expected dichroic ratios are listed in Table I for a number of values for θ [Eqs. (39) and (50),

TABLE I

EXPECTED DICHROIC RATIOS AS A FUNCTION OF THE ANGLE θ
BETWEEN THE TRANSITION MOMENT AND THE CHAIN AXIS FOR
PERFECT AXIAL ORIENTATION

Calculated from Eq. (39), Chapter V (orientation parameter, $S = 0$)

θ (deg)	R
90	∞
75	6.95
60	1.50
54.7	1
45	0.50
30	0.167
15	0.036
0	0

Chapter V]. It is interesting to note that for an angle $\theta = 54.7°$ no dichroism is expected even though the sample is perfectly oriented.

Finally, we want to mention that the knowledge of the dichroism may help in the assignment of absorption bands. This is particularly so for characteristic group vibrations where for an oriented sample it is often possible to predict whether a band should have parallel or perpendicular dichroism.

In Chapter V we will show in detail how to derive useful quantitative correlations between the infrared dichroism and the type and degree of orientation as well as the geometrical structure of the polymer chain. No further details are, therefore, given in this section except for the following illustration.

C. Illustration

We briefly discuss some aspects of the polarization spectra of a stretched film of polyacrylonitrile given in Fig. 2. The spectrum shows several major absorption bands; some are perpendicular, others are parallel bands. The dichroic ratio R varies over a wide range for the observed absorption bands. For this particular sample an X-ray diffraction pattern also showed that the chains are oriented to a high degree. The transition moment for the —C≡N stretching vibration and the —C≡N bond itself are more or less perpendicular to the chain axis since the corresponding absorption band at 2245 cm^{-1} has a dichroic ratio of 2.3. Observed dichroic ratios are shown in Table II (Ref. 10) for a series of polyacrylonitrile

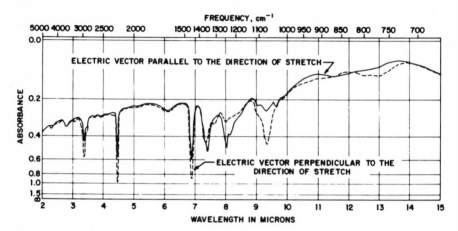

Fig. 2. Polarization spectra of polyacrylonitrile film (0.030 mm). Key: ——: electric vector parallel to the direction of stretch; – – –: electric vector perpendicular to the direction of stretch.

fibers drawn by various amounts. In the same table the orientation angle determined by wide-angle X-ray diffraction is listed (180° and 0° correspond to random and perfect orientation, respectively). From this table it is clear that the dichroic ratio can be used as a measurement of the degree of orientation.

An interesting observation is worth mentioning in connection with orientation measurements on polyacrylonitrile. If a sample, containing a small amount (a few percent) of residual solvent, is stretched, it is possible that not only the polymer but also the solvent will be oriented. In a particular experiment by Bohn (Ref. 9) a fiber of polyacrylonitrile spun from a dimethylformamide solution was stretched 4× in hot water. Orientation measurements on this fiber in an infrared microscope (see Chapter V) showed that the trace of solvent left in the fiber was moderately oriented. The dichroic ratio for the —C≡N absorption band at 2245 cm^{-1} was 2.30 ± 0.12 while for the C=O stretching absorption band of dimethylformamide at about 1660 cm^{-1} it was 1.44 ± 0.05. This finding indicates that a specific polymer-solvent interaction tends to align the solvent molecules with the C=O bond preferentially perpendicular to the axis of the polymer chain. Such orientation measurements on other systems might yield valuable information about the mechanism by which a solvent molecule interacts with a polymer chain.

TABLE II

INFRARED DICHROIC RATIO FOR THE —C≡N STRETCHING VIBRATION
AND X-RAY DIFFRACTION ORIENTATION ANGLE FOR POLYACRYLONITRILE
FIBERS DRAWN BY VARIOUS AMOUNTS (Refs. 9, 10)

Draw ratio	Dichroic ratio of —C≡N band	X-ray orientation angle (deg)
As spun[a]	1.09 ± 0.03	120 ± 10
2.5 ×	1.71 ± 0.04	36 ± 1
4 ×	2.00 ± 0.11	22 ± 1
6 ×	2.40 ± 0.05	21 ± 1
8 ×	2.66 ± 0.10	19 ± 1
10 ×	2.80 ± 0.08	17 ± 1
16 ×	3.11 ± 0.17	15 ± 1

[a] This particular fiber was slightly oriented in the spinning process.

4. Crystallinity

A. INTRODUCTION

In Section 1 we have discussed the type of crystallinity characteristic for polymers. We particularly have pointed out that in most cases the crystallinity is not perfect and that a sample will have amorphous as well as crystalline regions. These regions are expected to have different absorption spectra, but in an actual experiment it is not possible to observe the two types of spectra separately since the radiation beam in a spectrophotometer always covers an area much larger than the cross section of an individual crystalline region. Nevertheless it is possible to determine the effects of crystallinity on the infrared absorption spectrum by comparing a fully amorphous with a highly crystalline sample of the same polymer (Ref. 53). Observed differences will be described in the following paragraphs for a few examples. An interpretation of such differences and a summary of the crystallinity effects will be given at the end of this section. We also will discuss how some spectra can be used to determine the percent crystallinity in a polymer sample.

B. Discussion of Observed Spectra

Heneicosanoic Acid

As a first example we consider the spectrum of Heneicosanoic acid which is a relatively low molecular weight chain molecule. Figure 1 shows the spectra of the compound in the crystalline state and as a solution (amorphous state) in carbon disulfide. The most striking differences between the two spectra are observed in the 1200–1300 cm^{-1} region and in the 725 cm^{-1} region. In the crystalline state the CH$_2$-twisting absorption is split into nine peaks while the CH$_2$-rocking band is split into a doublet. The latter splitting is characteristic for most compounds containing a linear chain of CH$_2$-groups and in particular for polyethylene. It is caused by a well-defined intermolecular interaction. The band is a doublet because there are two chains passing through a crystallographic unit cell. Further details on this subject are given in Chapter II with a discussion of the spectrum of polyethylene. In Chapter IV we will show that the splitting of the CH$_2$-twisting absorption is caused by an intramolecular interaction of the 19 CH$_2$-groups, which in this particular case can be treated as a set of 19 coupled oscillators. The number of the observed absorption peaks, therefore, depends upon the number of CH$_2$ groups in the chain. The splitting of the CH$_2$-twisting absorption into nine peaks for a crystalline sample is only indirectly due to the crystalline state since it is not caused by crystal lattice interactions, but rather by the fact that we have a perfectly straight zig-zag chain, and it is only in the crystalline state that a perfectly straight chain is thermodynamically stable. In solution the CH$_2$-groups no longer form a linear zig-zag chain, since rotational isomers can occur in a more or less random fashion (Ref. 57). Therefore, the CH$_2$ groups cannot interact in a regular manner and only a very broad and rather poorly defined absorption band can be observed. A crystalline chain consisting of a different number of CH$_2$ groups would have a distinctly different absorption spectrum in the region of the CH$_2$-twisting vibrations. Aronovic (Ref. 1) has made use of these differences to distinguish between very similar fatty acids. For example the spectra of CH$_3$—(CH$_2$)$_{26}$—COOH and CH$_3$—(CH$_2$)$_{28}$—COOH are quite different between 1170 cm^{-1} and 1350 cm^{-1} (see also Chapter IV).

Poly(ethylene terephthalate)

Another case where the major spectral differences between crystalline and amorphous samples are caused by rotational isomerism rather than by differences in intermolecular interactions is poly(ethylene terephthalate). The solid line in Fig. 3 is a spectrum of an amorphous film prepared by rapidly quenching the molten polymer. The dashed line is a spectrum of the same film after crystallization by holding the sample at 160° C for several hours. Ward (Ref. 84, 85) as well as Daniels and Kitson (private communication, see also Ref. 25) have suggested that most of the crystallinity bands are due to the $\diagup$O$\diagdown$CH$_2$$\diagup$CH$_2$$\diagdown$O$\diagup$ group in the *trans*-form while some of the amorphous bands can be assigned to the *gauche*-form of the same group. The *gauche*-isomer is obtained from the *trans*-form by a rotation by 120° around the C—C axis as illustrated in Fig. 4. To support this hypothesis the authors (mentioned above) have compared the spectra of amorphous and partially crystallized poly(ethylene terephthalate) with the spectra of the following

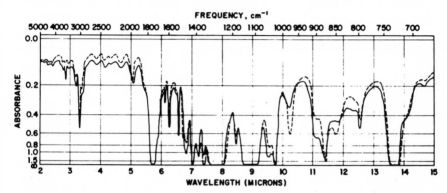

FIG. 3. Spectrum of unoriented film of poly(ethylene terephthalate) (0.025 mm). KEY: ——: quenched from the melt; ---: crystallized at 160° C.

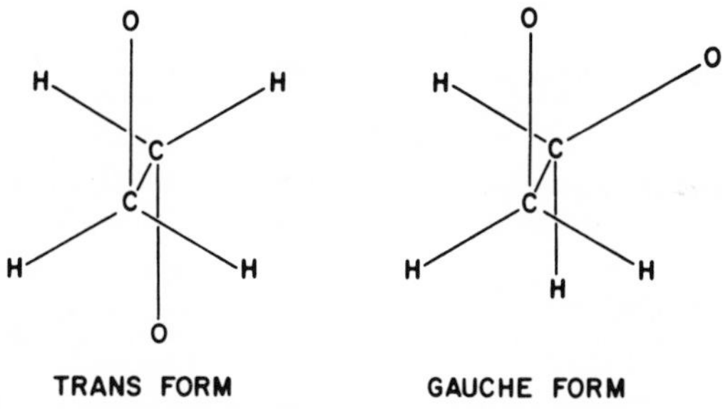

FIG. 4. *Trans*-form and *gauche*-form for the $\diagdown$O$\diagup$CH$_2\diagdown$CH$_2\diagup$O$\diagdown$ group in poly(ethylene terephthalate).

three model compounds. The first one is a cyclic trimer of poly(ethylene terephthalate). It can be obtained as a single crystal and molecular models suggest that all three $\diagup$O$\diagdown$CH$_2$CH$_2\diagdown$O$\diagup$ groups are in the *gauche*-configuration. The second one is the linear model compound

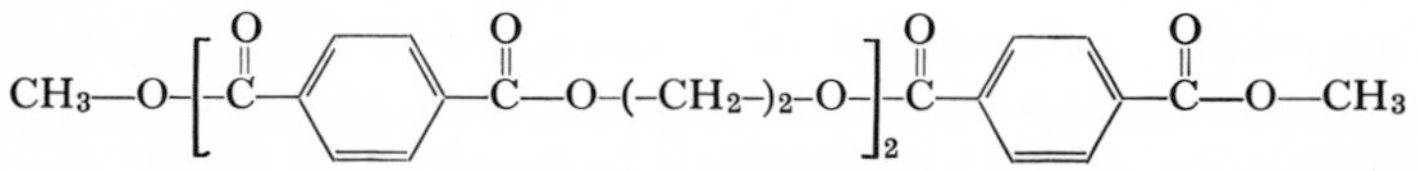

$$CH_3-O-\left[\overset{O}{\overset{\|}{C}}-\bigcirc-\overset{O}{\overset{\|}{C}}-O-(-CH_2-)_2-O-\right]_2\overset{O}{\overset{\|}{C}}-\bigcirc-\overset{O}{\overset{\|}{C}}-O-CH_3$$

which has a crystal structure isomorphous with the one of the polymer and, therefore, is expected to have the *trans*-configuration (Ref. 26). The third one is ethylenedibenzoate for which the spectrum was obtained in the solid state and as a dilute solution in carbon tetrachloride. The major bands sensitive to configurational changes are listed in Table III. It is important to note that the amorphous polymer contains bands due to the *trans*- as well as the *gauche*-form. Therefore, both configurations are present in the amorphous sample. When such a sample is crystallized some of the aliphatic groups in the *gauche*-form are converted into the *trans*-form with a decrease in the absorption intensity at 1445,

TABLE III

ABSORPTION FREQUENCIES IN CM^{-1} FOR THE *gauche*- AND *trans*-FORMS OF THE —O—CH₂—CH₂—O— GROUP IN POLY(ETHYLENE TEREPHTHALATE) AND RELATED COMPOUNDS

	CH₂		C—O	
Compound	Bending	Wagging	Stretching	Others
Cyclic trimer (Ref. 85), *gauche*-form	1445	1370		1045, 900
Linear model compound (Ref. 85) *trans*-form	1470	1340		975
Ethylenedibenzoate, crystalline (Ref. 25) *trans*-form	1480	1335	1115	978
Ethylenedibenzoate, dilute solution in CCl₄ (Ref. 25), mostly *gauche*-form	1480 (trace)	1370	1097	
Poly(ethylene terephthalate) (see Fig. 3) "crystallinity bands" (*trans*-form)	1470 1473†	1340 1342†	1120 1128†	975, 850 972†, 850†
"amorphous bands" (*gauche*-form)	1445	1370 1372†	1110 1100†	1045, 900 1043†
(Ref. 85, the values with a dagger are from Ref. 25)				

1370, 1110, 1045, and 900 cm^{-1} and an increase in intensity of the corresponding *trans*-bands. Such spectral changes can be used to determine quantitatively the changes in percent crystallinity. But a word of caution is in order. Since the *trans*-bands are not true "crystallinity bands" their intensity can be high for a sample with relatively low crystallinity if the *trans* content for the amorphous fraction of the sample is high. This may particularly be true for a highly oriented sample with rather extended polymer chains. In such a case a percent crystallinity determination based on the *trans*-band intensities tends to result in values that are too high (Refs. 30, 31).

In another application Cobbs and Burton (Ref. 18) have used the 972 cm⁻¹ band to measure crystallization rates. They mounted amorphous films in a heated infrared absorption cell and followed the intensity increase as a function of time and temperature.

Polypropylene

Polypropylene is probably the polymer that shows the most striking crystallinity effects in the infrared spectrum. Figure 5 gives a spectrum of a highly crystalline and a molten sample. Some bands present in the spectrum of the crystalline

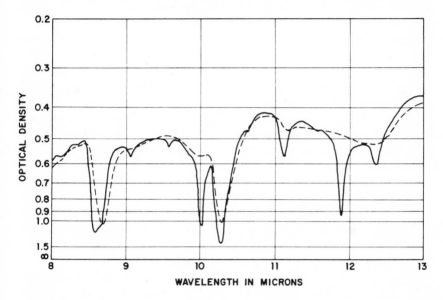

Fig. 5. Spectrum of isotactic polypropylene in the 8–13 μ region. KEY: ——— : film 0.076 mm thick at room temperature; ——— : same film molten between two salt plates at 240° C.

sample are completely missing in the spectrum of the molten sample. The band in the 8.6 μ region is shifted and broadened by going from the crystalline to the amorphous state, and in the 12 μ region only a very broad absorption is observed in the melt. These spectral differences can be used to determine the percent crystallinity of polypropylene by measuring the relative absorbance of a crystallinity band and a crystallinity insensitive band. This measurement alone cannot lead to any absolute value for the percent crystallinity. To obtain such values one has to calibrate the infrared measurements with absolute values obtained by an X-ray method, like the one proposed by Natta, Corradini, and Cesari (Ref. 60) or by a density determination used by Quynn, Riley, Young, and Noether (Ref. 67). If we assume a density at 25° C of 0.856 g/cc and 0.936 g/cc for a fully amorphous and a 100% crystalline sample, respectively (Ref. 59), and measure the intensity of the crystallinity sensitive band at 10.03 μ and the amorphous band at 10.29 μ, the following linear equation will express the percent crystallinity (Ref. 67):

$$\% \text{ crystallinity} = 109 \frac{A(10.03\ \mu) - A(10.90\ \mu)}{A(10.29\ \mu) - A(10.90\ \mu)} - 31.4$$

$A(\lambda)$ is the optical density at the wavelength λ. 10.90 μ is used as a convenient reference point in the determination of the intensity of the two absorption bands. Such an infrared method has the advantage of being fast, simple, and more sensitive to small changes in crystallinity than the X-ray method (Ref. 60).

6-6 Nylon

An example where it is possible to obtain absolute values for the percent crystallinity by an infrared method alone is the case of 6-6 nylon. This spectrum has separate absorption bands that are due only to amorphous and crystalline material respectively. Starkweather and Moynihan (Ref. 76) have chosen the amorphous band at 8.8 μ and the crystallinity band at 10.7 μ for their analysis. The authors measured the absorption intensity of the two bands for a large number of samples of various degree of crystallinity. They plotted the band intensity for both bands as a function of the density for each sample, since the density can be used as a relative measure for the amount of crystallinity. They found a linear relationship as one would expect for pure crystallinity and pure amorphous bands, where the absorption intensity should be proportional to the relative amount of the respective components. A linear relationship is also expected between the density and the percent crystallinity as pointed out in Section 1. Samples of a 100% amorphous or a 100% crystalline polymer were not available but an absolute crystallinity scale can still be obtained by extrapolating the straight lines in Fig. 6 to zero absorbance for the 8.8 μ and 10.7 μ bands. With the help of this plot the percent crystallinity can be determined in any 6-6 nylon film by just measuring the film thickness and the intensity of one of the two bands. Figure 6 also gives the density for a fully amorphous and 100% crystalline sample. The values are 1.069 ± 0.002 and 1.220 ± 0.002 g/cc, respectively. By using these two points in a straight line plot a simple density determination can give an independent check for the percent crystallinity.

C. INTERPRETATION AND SUMMARY

In this subsection we will summarize some of the reasons for differences in the spectra of polymers in the amorphous and crystalline state. These differences can be classified into two groups. They are caused by variations in either intramolecular or intermolecular interactions. They will be discussed in this order in the following two paragraphs.

We describe the effects of intramolecular interactions for an individual polymer

chain. In a first approximation such a chain in the crystalline state can be considered a one-dimensional infinite crystal with the polymer repeat unit being the
unit cell. In Chapter II we will show that the spectrum for a straight infinite chain
is expected to be quite simple because of very strict selection rules. For relatively
short chains of some linear hydrocarbons, however, these selection rules can
break down for certain characteristic group vibrations and a splitting of an

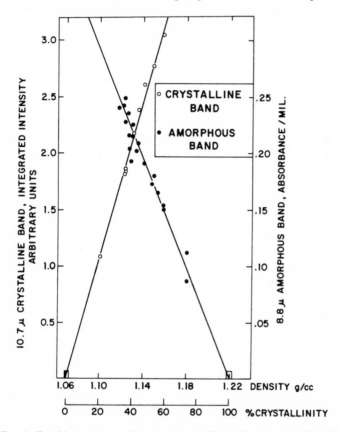

FIG. 6. Band intensity as a function of crystallinity for 6-6 nylon (Ref. 76).

absorption into many components can be observed (Fig. 1). For the amorphous
state of polymers where the chains can assume the form of random coils the strict
selection rules do not hold either. Transitions forbidden in the crystalline state
may, therefore, be allowed for amorphous chains, resulting in absorption bands
characteristic for the amorphous state only. The intramolecular interactions in a
straight chain are quite regular, leading to sharp and well-defined absorption
bands. For randomly coiled chains these interactions are irregular, which causes
a broadening of bands. In some polymers such random coils are formed by a

random distribution of two or more well-defined rotational isomers of certain segments along the chain. An example is poly(ethylene terephthalate) where in the crystalline state the two adjacent CH_2-groups are in the *trans*-position while in an amorphous sample the CH_2-groups predominantly assume the *gauche*-configuration (Fig. 4). The spectra of two such isomers are distinctly different (Fig. 3). The differences discussed in this paragraph are caused by the fact that a chain assumes a different configuration in the amorphous and crystalline state. Other differences due to intermolecular interactions are summarized in the following paragraph.

The interaction between adjacent chains in a polymeric crystalline region causes absorption bands to split. The number of peaks in the split band can be as high as the number of chains passing through a unit cell. For example, for polyethylene this number is two (see Chapter II) causing a splitting of the CH_2-bending and -rocking vibrations into doublets. If for such a split band only one of the two components is infrared-active the band for a crystalline sample may appear to be shifted compared to its position for the amorphous state. A broadening of absorption bands in the amorphous state occurs for most compounds because of the random type of intermolecular interaction that produces a slightly different force field for each absorbing group. Finally we point out that in some cases a polymer can assume different crystalline forms. The spectra of these forms are also different even though the actual configuration of an individual chain may be the same.

5. Analytical Applications

A. Introduction

The analytical aspect of infrared spectroscopy is of great practical importance to the chemist. For the last fifteen years he has been using this tool extensively for qualitative and quantitative analysis, particularly in the field of organic chemistry. It is not our intention to treat this subject in a general way since a number of textbooks and handbook articles (see, e.g., Refs. 13, 19, 23, 45, 46, 50, 55, 68) were written for this purpose. Bellamy's book (Ref. 4) is the most comprehensive one. At this point we will restrict ourselves to problems that are of interest to polymer science.

Characteristic absorption bands caused by group vibrations occur in a polymer spectrum just as they do in the spectrum of a nonpolymeric molecule as discussed in Section 2. These bands can, therefore, be used to identify the presence and determine the relative concentration of characteristic groups such as $>C=O$, $>CH_2$, phenyl, etc. On the other hand, there are bands due to vibrations involving the whole chain or chain segments. These bands are characteristic for the polymer as a whole. They are, therefore, used for the purpose of identification of the spectrum of an unknown polymer with a known spectrum (Refs. 35, 42, 43, 44, 47, 62, 73). In the following subsections a number of examples and different types of applications will be discussed.

B. Conventional Type of Infrared Analysis

The kind and amount of impurities, residual solvent, plasticizer, etc., present in a polymer can often be determined if the spectrum of the pure polymer is known, and if the compound to be determined has absorption bands that are not masked

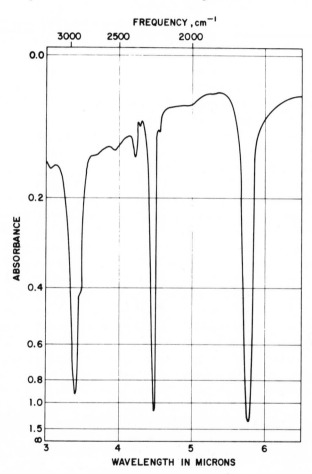

Fig. 7. Infrared spectrum of acrylonitrile/methacrylate (94/6) copolymer in the 3–6.5 μ region. Nitrile band at 4.47 μ and carbonyl band at 5.77 μ.

by the spectrum of the polymer. Polymer blends and copolymers can also be analyzed in this fashion. As an example the spectrum of a copolymer of acrylonitrile with methyl acrylate is shown in Fig. 7. The ester carbonyl band at 5.77 μ in the spectrum of the copolymer is absent in the spectrum of pure polyacrylonitrile. The absorption intensity of this band is proportional to the amount of methyl acrylate present in a given sample, while the nitrile band at 4.47 μ is characteristic for pure polyacrylonitrile. The copolymer ratio C_M/C_A is proportional to the ratio of the two absorption band intensities:

$$\frac{C_M}{C_A} = k \frac{A_{5.77\,\mu}}{A_{4.47\,\mu}} \tag{1}$$

where C_M and C_A are the molar concentrations of methyl acrylate and acrylonitrile monomer, respectively. A is the absorbance for the corresponding wavelength. To determine the value of the constant k a sample of a known copolymer ratio is required (Ref. 56). For this specific case it turns out that $k \approx 0.05$. This value is only correct for certain experimental conditions, since the peak absorption intensity (per unit sample thickness) of the extremely sharp nitrile band depends somewhat on the resolving power of the spectrophotometer as well as the thickness of the sample. These effects may cause a variation in k of as much as 20%.

C. CHEMICAL REACTIONS INVOLVING POLYMERS

It is possible to follow chemical reactions by observing changes in the infrared spectrum. Some times the rate of a polymerization reaction can be measured by determining the loss of monomer and increase in polymer with the help of suitable absorption bands.

The absorption spectrum is often used to study thermal and photodegradation of polymers. In a typical experiment a film is exposed to heat or light (usually ultraviolet) for various lengths of time and the spectrum is recorded after each time interval. In this fashion rates of degradation as well as information about the reaction products can be obtained. Figure 8 shows the spectral changes in the 4.5–7.5 μ region during photooxidation of a film of polyethylene. The strong absorption band at 5.82 μ that appears on exposure is due to the $>C{=}O$ stretching vibration of a decomposition product.

Under favorable circumstances we can follow cross-linking reactions by changes in the spectrum, but infrared is usually not the best tool for this purpose since only a few cross-links between chains might change the physical properties (especially solubility) of the polymer radically and hardly influence the infrared spectrum.

D. END GROUP ANALYSIS

The knowledge of the amount and type of end groups in a polymer sample can be quite useful. For example, for vinyl polymers the structure of some of the end groups depends on the type of catalyst used for the polymerization so that an analysis of end groups can help in a study of the mechanism of polymerization. A long chain polymer has fewer end groups per unit weight than one consisting of short chains. The number average molecular weight can, therefore, be obtained by determining the end group concentration. Complications arise if a molecular chain is not linear but branched so that an individual molecule has more than two end groups. Other complications arise from the fact that many polymers have more than one kind of end group.

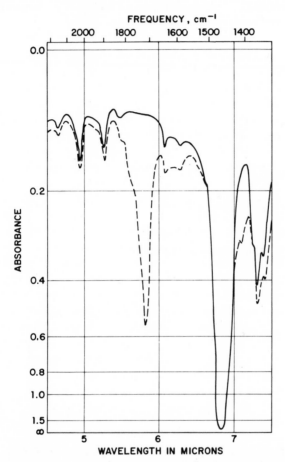

FIG. 8. Photooxidation of polyethylene. Infrared spectra in the 4.5–7.5 μ region. KEY: ———: unexposed film, 0.13 mm thick; – – –: exposed to UV light for two weeks in the presence of oxygen and water vapor.

A number of chemical techniques are known for end group analysis (see, e.g., Refs. 11, 66) but at this point we will restrict ourselves to a few examples where infrared methods can be used successfully.

The end groups in poly(ethylene terephthalate) are —CH_2—CH_2—OH. On heat degradation carboxyl-end groups are formed (Ref. 25). The R—OH group has a characteristic OH-stretching band at 2.83 μ while the COOH group has a strong and broad absorption in the 3.05 μ region. An intensity measurement of these bands can be used to determine the absolute concentration of the two types of end groups present in a given sample provided the absorbance has been calibrated for at least two samples with a known end group concentration. Figure 9 shows two spectra obtained by Daniels and Kitson (Ref. 25). The concentration of end groups for the two corresponding samples is also given in Fig. 9.

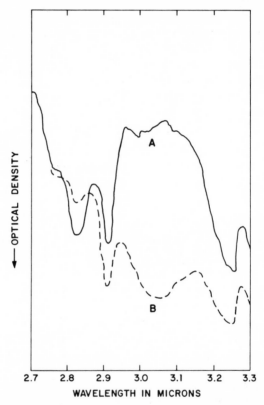

FIG. 9. Spectrum of poly(ethylene terephthalate) after Daniels and Kitson (Ref. 25). Carboxyl OH-stretching band at 3.05 μ. Carboxyl concentration: (A) 5.2 equivalents/10^6 grams; (B) 221 equivalents/10^6 grams.

Infrared spectroscopy has been used for the determination of the methyl group concentration in polyethylene (Refs. 12, 14, 69, 72). Methyl groups can occur only as end groups of chain molecules or as end groups of side branches. The determination of the methyl group concentration can, therefore, be an indication of the degree of branching, especially for samples of known molecular weight. This knowledge is of considerable practical importance since the physical properties of polyethylene depend strongly on the degree of linearity of the molecular chain. The symmetric C—H bending vibration band at 7.25 μ has been used most frequently for the methyl group analysis (Refs. 14, 69). Figure 10 shows two spectra of polyethylene with 1.6 and 34 methyl groups per 1000 carbon atoms, respectively. In a first approximation the absorbance of the 7.25 μ band is proportional to the methyl group concentration, and the following equation can be used

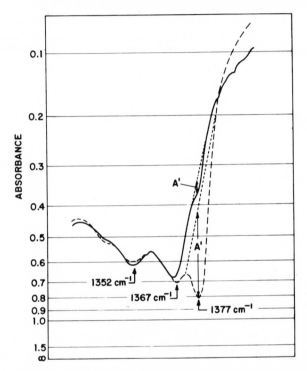

FIG. 10. Absorption spectra of polyethylene in the 1370 cm^{-1} region. The spectra were recorded on a Perkin-Elmer single beam double pass spectrophotometer equipped with a calcium fluoride prism. KEY: ———: film thickness 0.21 mm, 1.6 methyl groups/1000 carbon atoms; ———: film thickness 0.16 mm, 34 methyl groups/1000 carbon atoms. A' is the absorbance of the 1377 cm^{-1} band. It is determined from an arbitrarily chosen base line.

to determine the concentration from a simple intensity measurement. For the instrument described in Fig. 10 we obtain

$$N = 13.0A \qquad (2)$$

where N is the number of methyl groups per 1000 carbon atoms, and A is the absorbance (at room temperature) for a film 1 mm thick. Reding and Lovell (Ref. 69) and Boyd, Voter, and Bryant (Ref. 12) found that an equation such as Eq. (2) can be only an approximation since the extinction coefficient varies for different types of methyl groups, that is, methyl groups attached to side branches of a different length. By measuring the intensity of the 7.25 μ band for polymers such as polypropylene, poly-1-butene, poly-1-heptene, etc., they found a spread in extinction coefficient of almost a factor of two. Without the knowledge of the length of the side branches for a sample of polyethylene the 7.25 μ band, therefore, cannot be used to determine accurately the concentration of methyl groups. The authors found that the length of a short side branch can be determined from the position of the methyl rocking vibration in the 11 μ region. They found, for example, that this band occurs at 10.28 μ for methyl branches, at 11.18 μ for butyl branches, and at 11.25 μ for a long side branch. The reason for such variations is discussed in Chapter IV, Section 4B. From an analysis of these bands† it turns out that for a variety of polyethylene samples only two kinds of side branches are present, namely, ethyl branches and butyl branches. Furthermore, the ratio of ethyl groups to butyl groups is found to be roughly constant and equal to about 0.6 to 0.7 for a variety of samples. For almost linear polyethylene very few methyl groups are present and the methyl rocking vibration bands in the 11 μ region are too weak to be observed. The 7.25 μ band then remains the only indication of methyl groups either in side branches or as end groups of the chains. For more detailed information on this subject the reader should refer to Ref. 12.

E. UNSATURATIONS

In the footnote of the previous paragraph we have indicated that irregularities in a polymer chain such as unsaturations can be identified by characteristic absorption bands. With the help of the C—H out-of-plane bending vibrational bands it is possible to distinguish between different kinds of unsaturations, and from a measurement of the band intensities one can compute the relative abandonce of the groups. Table IV is a summary of literature data (Refs. 4, 7, 8, 24, 27, 32, 70, 82) giving frequency and band intensity of the C=C stretching vibrations and the C—H out-of-plane bending vibrations for various types of double bonds.

An analysis of unsaturations in polymers is of great practical importance in a number of cases. Polyethylene, for example, can contain various types of double bonds. The knowledge of type and amount of these double bonds is particularly useful in elucidating the polymerization mechanism. Butadiene (CH_2=CH—CH=CH_2) can be polymerized by a 1,2-addition, a *cis*- or *trans*-1,4-addition or as a mixture of all three depending upon the type of catalyst used. The relative amount of each type can be determined by a simple measurement of the corresponding band intensities. Polyisoprene is a case of particular practical importance. The polymerization of isoprene

$$\underset{1}{CH_2}=\underset{2}{\overset{\overset{\textstyle CH_3}{|}}{C}}\text{——}\underset{3}{\overset{\overset{\textstyle H}{|}}{C}}=\underset{4}{CH_2}$$

† Vinyl and vinylidene groups (present as impurities) absorb at 11.0 and 11.27 μ, respectively, interfering with the methyl absorption. These groups have, therefore, to be removed by hydrogenation of the sample before analysis.

to natural rubber which has the pure 1,4-structure

$$-(CH_2 \overset{CH_3}{\underset{}{}} C=C \overset{H}{\underset{CH_2}{}})_x$$

has been a challenge to polymer chemists for many years. With the help of co-ordination catalysis it was possible to produce synthetic "natural rubber" (Ref. 40, 77). The pure *cis*-1,4-structure was confirmed by infrared spectroscopy. From

TABLE IV

FREQUENCY AND INTENSITY OF ABSORPTION BANDS CHARACTERISTIC FOR VARIOUS KINDS
OF DOUBLE BONDS[a]

	C=C Stretching		C—H Out-of-plane bending	
	ν (cm^{-1})	ϵ^b	ν (cm^{-1})	ϵ^b
R—CH=CH₂, two C—H bands {	1642 ± 3	28–44	990–1003	33–57
			908–916	110–150
1,2-polyisoprene			909	149
1,2-polybutadiene			911	145
R—CH=CH—R′ (*trans*)	1667	Very weak	964–977	85–120
trans-1,4-polybutadiene			967	109
R—CH=CH—R′ (*cis*)	1646 ± 10	7–18	675–729	13–106
cis-1,4-polybutadiene			680	23
RR′C=CH₂	1650 ± 11	20–42	883–895	103–200
3,4-polyisoprene			888	145–159
RR′C=CHR″	1680 ± 10	4–7	788–840	11–35
cis-1,4-polyisoprene			835	19
trans-1,4-polyisoprene			842	11

[a] The data represent a summary of literature values from Refs. 4, 7, 8, 24, 27, 32, 70, 82.
[b] The molar extinction coefficient ϵ is defined by the following equation:

$$\epsilon = \frac{M}{cd} \log \frac{I_0}{I}$$

where M is the molecular weight of absorbing group (i.e., 24 for C=C), c is the concentration of the group in g/liter, and d is the cell thickness in cm.

Table IV we find that 1,2- and 3,4-additions can easily be identified by their absorption bands at 909 and 888 cm^{-1}, respectively, and relatively small amounts of these structures can be detected. It is much more difficult to distinguish quantitatively between the *cis*-1,4- and the *trans*-1,4-configuration since the respective C—H out-of-plane bending vibration bands are fairly weak and of almost the same frequency as shown in Fig. 11. But Richardson and Sacher (Ref. 70) have worked out a method for determining the relative amount of the four

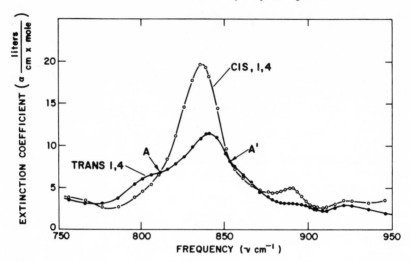

Fig. 11. Absorption spectra of *cis*-1,4- and *trans*-1,4-polyisoprene, after Richardson and Sacher (Ref. 70).

different isomers by measuring the extinction coefficient at five different frequencies (only four would be necessary). Their results are listed in Table V. The analysis is based on the assumption that the observed extinction coefficient ϵ_ν at a certain frequency ν is given by the following equation:

$$\epsilon_\nu = \frac{\epsilon_{1\nu}c_1 + \epsilon_{2\nu}c_2 + \epsilon_{3\nu}c_3 + \epsilon_{4\nu}c_4}{c_1 + c_2 + c_3 + c_4} \tag{3}$$

where $\epsilon_{1\nu}, \ldots, \epsilon_{4\nu}$ are the extinction coefficients for the four isomers at the frequency ν and $c_1, \ldots, c_4$ are the concentrations of the four isomers. Similar equations

TABLE V

Extinction Coefficient ϵ for the Four Isomers of Polyisoprene after Richardson and Sacher (Ref. 70)

Structure	ν (cm^{-1})				
	815.5	843.0	857.5	887.5	909.0
1,2-Addition	0.31[a]	0.62[a]	1.02[a]	5.73[a]	149
3,4-Addition	0.69[a]	1.80[a]	3.87[a]	159	7.48[a]
cis-1,4-Addition	6.78	19.2	8.06	4.56	2.72
trans-1,4-Addition	6.78	11.3	8.06	3.26	2.40

[a] Olefinic model compounds do not show an absorption at these positions. The values are, therefore, only accurate for a polyisoprene analysis.

hold for every frequency, and if we measure ϵ_ν at four of the five frequencies (listed in Table V) we obtain four equations which can be solved for the four unknown concentrations $c_1, \ldots, c_4$. This is not the only choice of frequencies for solving this problem. Binder and Ransaw (Ref. 8) have, in fact, shown that the analysis may be more accurate for certain concentration ranges by a somewhat different choice of analytical frequencies.

F. Configurational Isomers

Configurational isomers of a polymer have usually different absorption spectra. Such differences can often conveniently be used to distinguish between isomers

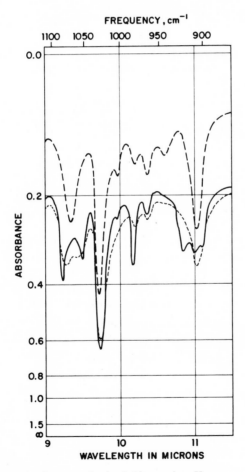

Fig. 12. Spectra of polystyrene in the 9–12 μ region. Key: ———: isotactic crystalline film; ---: isotactic crystalline film above the crystalline melting point; — — —: atactic amorphous film.

and to analyze an unknown mixture in a fashion similar to the one just described. A number of examples will be mentioned in the following paragraphs.

There are characteristic differences in the spectra of amorphous isotactic and atactic polystyrene. These differences are even more pronounced if the isotactic

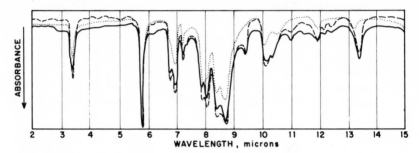

FIG. 13. Spectra of various poly(methyl methacrylate) (Ref. 3). KEY: ——: atactic; ....: isotactic; ———: syndiotactic.

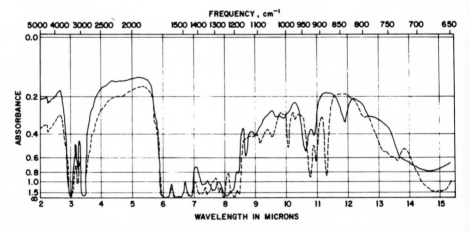

FIG. 14. Poly(hexamethylene hexahydroterephthalamide)

KEY: ——: *cis*-isomer, film 0.038 mm thick; – – –: *trans*-isomer, film 0.054 mm thick.

form is crystallized as shown in Fig. 12. Similar, but less pronounced, differences are also observed between the spectra of conventional, syndiotactic, and isotactic poly(methyl methacrylate) as illustrated by the three spectra of Fig. 13 (Ref. 3). The purity of the isotactic form can easily be determined with the help of the 9.45 μ band which is absent for the pure isotactic polymer. For the atactic and syndiotactic polymer the observed differences are relatively small. In this case infrared spectroscopy is a rather insensitive tool to detect small variations in tacticity.

Cis- and *trans*-isomers of polymers can usually be distinguished quite readily by their infrared spectra. As an example the spectra of the two isomers of poly-(hexamethylene hexahydroterephthalamide) are shown in Fig. 14. Very pronounced differences are observed in the 700–1400 cm^{-1} region. By determining the intensity of absorption bands characteristic for each one of the isomers we can analyze a polymer mixture or a copolymer consisting of the two isomers. In this particular case the accuracy of such an analysis is quite high.

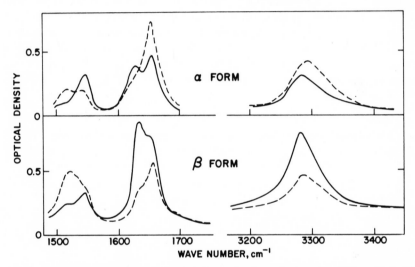

FIG. 15. Absorption spectra of rolled films of poly-L-alanine after Elliott (Ref. 28). KEY: ———: electric vector of radiation perpendicular to the direction of rolling; –––: electric vector of radiation parallel to the direction of rolling.

Polypeptides are another example where infrared spectroscopy is very useful to distinguish between configurational isomers. Most synthetic polypeptides are polyamides of the following form:

$$X-(-CHR-CO-NH-)_n-Y$$

X and Y are end groups and R can represent a variety of side chains. If, for example, R is a hydrogen atom the polymer is polyglycine. A list of other polypeptides is given by Bamford, Elliott, and Hanby (Ref. 2) in their book on synthetic polypeptides. They also give an excellent account of the infrared spectra of these polymers. In this paragraph we will discuss only how the spectrum is affected if the chain is converted from the α-form to the β-form. The changes are particularly pronounced for the absorption bands characteristic for the amide group. The α-form is a folded helix where the C=O and N—H bonds are approximately parallel to the axis of the helix while in the β-form the chain is extended to an almost planar zig-zag chain. In this case the C=O and N—H bonds are close to being perpendicular to the chain axis. The spectra of the two forms, therefore, differ in their dichroic behavior if oriented samples are analyzed with polarized radiation. Figure 15 shows the absorption spectra of poly-L-alanine after Elliott (Ref. 28). The sample for the upper spectrum was mostly in the α-form while for the lower spectrum the β-form was predominant. From these spectra it is clear that the C=O band (in the 1600 cm^{-1} region) and N—H stretching band (in the 3300 cm^{-1} region) are parallel bands for the α-form and perpendicular bands for the β-form, while this is the other way round for the amide II band (1500 cm^{-1} region). Furthermore, it should be noted that the absorption maxima occur at different frequencies for the two different forms. A survey of a large number of polypeptides has shown (Ref. 2) that the position of the amide absorption bands is quite characteristic for the two forms as shown in Table VI.

TABLE VI

FREQUENCY RANGE IN CM^{-1} OF CHARACTERISTIC AMIDE ABSORPTION
BANDS FOR POLYPEPTIDES

	N—H Stretching band	C=O Stretching band	Amide II band
α-Form	3292–3305	1648–1662	1531–1558
β-Form	3283–3301	1629–1635	1515–1526

We conclude that dichroic measurements are very useful for the analysis of oriented polypeptides, while for randomly oriented samples we have to rely entirely on an accurate measurement of absorption frequencies to distinguish between the α-form and the β-form. For further details the reader should refer to references 2 and 29.

REFERENCES

1. Aronovic, S. M., Ph.D. Thesis, University of Wisconsin, 1957.
2. Bamford, C. H., Hanby, W. E., and Elliott, A. E., "Synthetic Polypeptides." Academic Press, New York, 1956.
3. Baumann, U., Schreiber, H., and Tessmar, K., *Makromol. Chem.* **36**, 81 (1959).
4. Bellamy, L. J., "The Infra-red Spectra of Complex Molecules," 2nd ed. Wiley (Methuen), New York, 1958.
5. Bhagavantam, S., and Venkatarayudu, T., *Proc. Indian Acad. Sci.* **A9**, 224 (1939).
6. Billmeyer, F. W., Jr., "Textbook of Polymer Chemistry." Wiley (Interscience), New York, 1957.
7. Binder, J. L., *Anal. Chem.* **26**, 1877 (1954).

8. Binder, J. L., and Ransaw, H. C., *Anal. Chem.* **29**, 503 (1957).

9. Bohn, C. R., private communication, 1956.

10. Bohn, C. R., Schaefgen, J. R., and Statton, W. O., *J. Polymer. Sci.* **55**, 531 (1961).

11. Bonnar, R. U., Dimbat, M., and Stross, F. H., "Number-Average Molecular Weights." Wiley (Interscience), New York, 1958.

12. Boyd, D. R. J., Voter, R. C., and Bryant, W. M. D., Paper presented at 132nd meeting of the American Chemical Society, New York, 1957. Abstract p. 8T.

13. Brügel, W., "Einführung in die Ultrarotspektroskopie." Steinkopff, Darmstadt, 1954.

14. Bryant, W. M. D., and Voter, R. C., *J. Am. Chem. Soc.* **75**, 6113 (1953).

15. Campbell, T. W., and Haven, A. C., *J. Appl. Polymer Sci.* **1**, 73 (1959).

16. Charch, W. H., and Moseley, W. W., *Textile Res. J.* **29**, 525 (1959).

17. Charch, W. H., and Shivers, J. C., *Textile Res. J.* **29**, 536 (1959).

18. Cobbs, W. H., and Burton, R. L., *J. Polymer. Sci.* **10**, 275 (1953).

19. Colthup, N. B., *J. Opt. Soc. Am.* **40**, 397 (1950).

20. Couture, L., *Compt. Rend.* **218**, 669 (1944).

21. Couture, L., *Compt. Rend.* **220**, 87 (1945).

22. Couture, L., *Compt. Rend.* **222**, 495 (1945).

23. Cross, A. D., "Introduction to Practical Infrared Spectroscopy." Butterworths, London, 1960.

24. Cross, L. H., Richards, R. B., and Willis, H. A., *Discussions Faraday Soc.* **9**, 235 (1950).

25. Daniels, W. W., and Kitson, R. E., *J. Polymer Sci.* **33**, 161 (1958).

26. Daubeny, R., Bunn, C. W., and Brown, C. J., *Proc. Roy. Soc.* **A226**, 531 (1954).

27. Davison, W. H. T., *Chem. & Ind.* (*London*) p. 131 (1957).

28. Elliott, A., *Proc. Roy. Soc.* **A226**, 408 (1954).

29. Elliott, A., *Advan. Spectry.* **1**, 213 (1959).

30. Farrow, G., McIntosh, J., and Ward, I. M., *Makromol. Chem.* **38**, 147 (1960).

31. Farrow, G., and Ward, I. M., *Polymer* **1**, 330 (1960).

32. Field, J. E., Woodford, D. E., and Gehman, S. D., *J. Appl. Phys.* **17**, 386 (1946).

33. Fischer, E. W., *Z. Naturforsch.* **12a**, 753 (1957).

34. Flory, P. J., "Principles of Polymer Chemistry." Cornell Univ. Press, Ithaca, New York, 1953.

35. Henniker, J., *Ind. Plastiques Mod.* (*Paris*) **12**, 30 (1960).

36. Herzberg, G., "Infrared and Raman Spectra of Polyatomic Molecules." Van Nostrand, Princeton, New Jersey, 1945.

37. Hexter, R. M., *J. Mol. Spectr.* **3**, 67 (1959).

38. Hexter, R. M., *J. Chem. Phys.* **33**, 1833 (1960).

39. Hexter, R. M., and Dows, D. A., *J. Chem. Phys.* **25**, 504 (1956).

40. Horne, S. E., Kiehl, J. P., Shipman, J. J., Folt, V. L., Gibbs, C. F., and co-workers, *Ind. Eng. Chem.* **48**, 784 (1956).

41. Huggins, M. L., "Physical Chemistry of High Polymers." Wiley, New York, 1958.

42. Hummel, D., *Kunstoffe* **46**, 442 (1956).

43. Hummel, D., *Farbe Lack* **62**, 529 (1956).

44. Hummel, D., "Kunstoff-, Lack- und Gummi-Analyse." Carl Hauser, Munich, 1958.

45. Jones, R. N., "Infrared Spectra of Organic Compounds: Summary Charts of Principle Group Frequencies," Bulletin No. 6. National Research Council, Ottawa, 1959.

46. Jones, R. N., and Sandorfy, C., *in* "Techniques of Organic Chemistry" (A. Weissberger, ed.), Vol. 9, Chapter 4. Wiley (Interscience), New York, 1956.

47. Kagarise, R. E., and Weinbreger, L. A., "Infrared Spectra of Plastics and Resins," Report PB-111-438. Naval Research Laboratory, Washington, D.C., 1954.

48. Keller, A., *Phil. Mag.* [8] **2**, 1171 (1957).

49. Ketelaar, J. A. A., *Angew. Chem.* **72**, 386 (1960).

50. Lecomte, J., *in* "Handbuch der Physik" (S. Flügge, ed.), Vol. 26, p. 244, Springer, Berlin, 1958.

51. Liang, C. Y., Krimm, S., and Sutherland, G. B. B. M., *J. Chem. Phys.* **25**, 543 (1956).

52. Mark, H., Marvel, C. S., Melville, H. W., and Whitby, G. S., eds., "High Polymers," Vols. 1–12. Wiley (Interscience), New York, 1940–1962.

53. Markova, G. S., Sadovskaya, G. K., and Kargin, V. A., *Zh. Fiz. Khim.* **30**, 437 (1956).

54. Mathieu, J. P., "Spectres de Vibrations et Symétrie des Molécules et des Cristaux." Hermann, Paris, 1945.

55. Mecke, R., and Kerkhof, F., *in* "Landolt-Börnstein, Zahlenwerte und Funktionen, Atom- und Molekularphysik," Vol. I, Part 2, p. 226. Springer, Berlin, 1951.

56. Miyake, A., *J. Chem. Soc. Japan, Ind. Chem. Sect.* **62**, 1449 (1959).

57. Mizushima, S., "Structure of Molecules and Internal Rotation." Academic Press, New York, 1954.

58. Natta, G., *Angew. Chem.* **68**, 393 (1956).

59. Natta, G., *Ind. Plastiques Mod. (Paris)* **10**, 40 (1958).

60. Natta, G., Corradini, P., and Cesari, M., *Atti Accad. Nazl. Lincei, Rend., Classe Sci. Fis., Mat. Nat.*, **22**, 11 (1957).

61. Neuilly, M., *Compt. Rend.* **238**, 65 (1954).

62. O'Neill, L. A., and Cole, C. P., *J. Appl. Chem. (London)* **6**, 399 (1956).

63. Person, W. B., and Olsen, D. A., *J. Chem. Phys.* **32**, 1268 (1960).

64. Person, W. B., and Swenson, C. A., *J. Chem. Phys.* **33**, 233 (1960).

65. Pimentel, G. C., McClellan, A. L., Person, W. B., and Schnepp, O., *J. Chem. Phys.* **23**, 234 (1955).

66. Price, G. F., *in* "Techniques of Polymer Characterization" (P. W. Allen, ed.), p. 207. Academic Press, New York, 1959.

67. Quynn, R. G., Riley, J. L., Young, D. A., and Noether, H. D., *J. Appl. Polymer. Sci.* **2**, 166 (1959).

68. Randall, H. M., Fowler, R. G., Fuson, N., and Dangl, J. R., "Infrared Determination of Organic Structures." Van Nostrand, Princeton, New Jersey, 1949.

69. Reding, F. P., and Lovell, C. M., *J. Polymer Sci.* **21**, 157 (1956).

70. Richardson, W. S., and Sacher, A., *J. Polymer Sci.* **10**, 353 (1953).

71. Rigaux, C., *Compt. Rend.* **238**, 63 (1954).

72. Rugg, F. M., Smith, J. J., and Wartman, L. H., *J. Polymer Sci.* **11**, 1 (1953).

73. Sawyer, R., S.P.E. *(Soc. Plastics Engrs.) J.* **15**, 537 (1959).

74. Sinclair, R. G., McKay, A. F., and Jones, R. N., *J. Am. Chem. Soc.* **74**, 2570 (1952).

75. Sorenson, W. R., and Campbell, T. W., "Preparative Methods of Polymer Chemistry." Wiley (Interscience), New York, 1961.

76. Starkweather, H. W., and Moynihan, R. E., *J. Polymer Sci.* **22**, 363 (1956).

77. Stavely, F. W., and co-workers, *Ind. Eng. Chem.* **48**, 778 (1956).

78. Stuart, H. A., "Die Physik der Hochpolymeren," Vols. 1–4. Springer, Berlin, 1952–1956.

79. Swenson, C. A., Person, W. B., Dows, D. A., and Hexter, R. M., *J. Chem. Phys.* **31**, 1324 (1959).

80. Swenson, C. A., and Person, W. B., *J. Chem. Phys.* **33**, 56 (1960).

81. Szigeti, B., *Proc. Roy. Soc.* **A258**, 377 (1960).

82. Thompson, H. W., and Torkington, P., *Trans. Faraday Soc.* **41**, 246 (1945).

83. Till, P. H., *J. Polymer Sci.* **24**, 301 (1957).

84. Ward, I. M., *Chem. & Ind. (London)* p. 905 (1956).

85. Ward, I. M., *Chem. & Ind. (London)* p. 1102 (1957).

86. Zbinden, R., *Helv. Phys. Acta* **26**, 129 (1953).

87. Zwerdling, S., and Halford, R. S., *J. Chem. Phys.* **23**, 2221 (1955).

—II—

Selection Rules for Chain Molecules

1. Introduction

In this chapter we will discuss the selection rules for crystalline systems (Refs. 7, 8, 13, 37) with emphasis on the special case of chain molecules (Refs. 12, 19). Such a discussion will lead to a basic understanding of the observed spectra.

The infrared and Raman spectra of any system are determined by its normal vibrations and by a set of selection rules. The number of normal vibrations in an N-atomic system is $3N$ (including translations and rotations of the system as a whole), while the number of infrared- or Raman-active vibrations may be much smaller. The latter can be determined by an analysis of the symmetry properties of the system, and in many cases it is possible to assign the observed frequencies to their corresponding normal vibrations. Such an analysis is usually quite simple for small molecules with a high degree of symmetry, but for large molecules without any symmetry this task is virtually impossible. For crystalline systems the problem becomes simple again because of the so-called translational symmetry. The group theoretical analysis of such systems is the subject of this chapter. We first will present the elements of group theory as far as they are needed in this book. Then we will briefly describe the well-known group theoretical treatment of simple polyatomic molecules and extend this treatment to crystalline systems and in particular to chain molecules.

2. Fundamentals of Group Theory

In this section, which is meant to be a brief summary, we will present the fundamentals of group theory. All those theorems and definitions will be given that are required for the understanding of the following sections. A number of good textbooks are available for a more detailed study of group theory (see, e.g., Refs. 3, 32, 36). Wigner's book (Ref. 36) is particularly useful as a guide to the theory of representations.

A. DEFINITIONS AND NOMENCLATURE

Group: A group consists of a number of abstract entities called "elements." A law of combination is defined so that two elements combined will result in an element of the group itself. Such association of two elements is called a "product." This product formation is associative, which means that the following three products of the three group elements A, B, C are identical:

$$(ABC) = A(BC) = (AB)C \tag{1}$$

A product is not necessarily commutative, so that in general

$$AB \neq BA \tag{2}$$

but groups that are commutative are called "Abelian." Furthermore, a group has always one and only one unit element E so that

$$ER = RE = R \tag{3}$$

where R can be any element of the group. Each element has an inverse element R^{-1} which is defined by the relation

$$RR^{-1} = R^{-1}R = E \tag{4}$$

The number of elements h is called the "order" of the group.

Symmetry Group: In the applications of group theory to spectroscopy we will only be concerned with "symmetry groups," where the elements are no longer abstract entities but rather symmetry operations applied to specific molecules or crystals.

Subgroup: It is possible that only part of the group elements form a group which is called a "subgroup" of the original group. The order g of the subgroup is always a divisor of h:

$$\frac{h}{g} = l \tag{5}$$

l, an integer number, is the "index" of the subgroup.

Isomorphous Groups: Two groups are isomorphous if every element in one group corresponds to an element in the other group and vice versa. Sometimes two groups are considered isomorphous if there is not a one-to-one correspondence of elements but instead of this, there may be several elements in one group corresponding to only one element in the other group. Some authors prefer to call this type of correspondence "homomorphism."

Conjugate Element: An element conjugate to R is XRX^{-1}, where X can be any element of the group.

Class: A group can be subdivided into classes. Elements in the same class are always conjugate to each other. If, therefore, one element of a class is given, all the others can be obtained by conjugation with every element of the group.

In an Abelian group every element forms a class by itself since

$$XRX^{-1} = XX^{-1}R = R \tag{6}$$

Self-Conjugate Subgroup (often called invariant subgroup): This is a subgroup which consists of full classes. A self-conjugate subgroup, usually denoted by $\mathfrak{N}$, consists of the n elements

$$E, \quad N_2, \quad N_3, \ldots, N_n \tag{7}$$

(This notation stems from the German word "Normalteiler.")

Coset: A coset is a set of elements obtained by multiplying (left or right) each element of a subgroup with one element of the group. The cosets of a self-conjugate subgroup are of particular importance. The left and right cosets are:

$$\begin{aligned} X\,\mathfrak{N} &= X, \quad XN_2, \quad XN_3, \ldots, XN_n \\ \mathfrak{N}X &= X, \quad N_2 X, \quad N_3 X, \ldots, N_n X \end{aligned} \tag{8}$$

The left and right cosets are identical as complexes:

$$X\,\mathfrak{N} = \mathfrak{N}X \tag{9}$$

but this identity does not hold in general for the individual elements

$$XN_\nu \neq N_\nu X \tag{10}$$

Factor Group: The elements of a factor group F are the cosets of a self-conjugate subgroup:

$$F = \mathfrak{N}, \quad \mathfrak{N}X_2, \quad \mathfrak{N}X_3, \ldots, \mathfrak{N}X_l = F_1, \quad F_2, \quad F_3, \ldots, F_l \tag{11}$$

A few explanatory remarks are in order. Each element of the factor group is the whole complex and not an individual element of the coset. The product of two factor group elements F_α and F_β is the coset F_γ which contains all the products of any group element of F_α with any group element of F_β. The unit element of the factor group is the self-conjugate subgroup itself. The order of F is equal to the number of nonequivalent cosets of $\mathfrak{N}$, and this is equal to the index l of $\mathfrak{N}$:

$$l = \frac{h}{n} \tag{12}$$

where h and n are the orders of the group and the self-conjugate subgroup, respectively.

B. Representation of a Group by Matrices

General Principle

Let us assume that a group consists of a number of abstract elements E, A, $B,\ldots, R,\ldots$, etc. It is possible to associate with each element R a square matrix $D(R)$. If such a set of matrices $D(E)$, $D(A)$, $D(B),\ldots, D(R),\ldots$ forms a group isomorphous with the group E, A, $B,\ldots, R,\ldots$, it is called a "representation" of the group (the conventional nomenclature D comes from the German word "Darstellung"). $D(R)$ is a matrix of the following form:

$$D(R) = \begin{pmatrix} D(R)_{11} & D(R)_{12}\ldots D(R)_{1n} \\ D(R)_{21} & D(R)_{22}\ldots D(R)_{2n} \\ \vdots & \\ D(R)_{n1} & D(R)_{n2}\ldots D(R)_{nn} \end{pmatrix} \tag{13}$$

The characteristic group relations which hold for the abstract elements also have to be fulfilled for the corresponding elements of the representation. If, for example,

$$AB = C \quad \text{then} \quad D(A)\,D(B) = D(C) \tag{14}$$

where the product of the two matrices $D(A)$ and $D(B)$ is formed according to the matrix multiplication rules

$$\sum_{l=1}^{n} D(A)_{il}D(B)_{lk} = D(C)_{ik} \tag{15}$$

Detailed information on matrices can be found in many textbooks on vector and tensor analysis. A short but very good introduction to this subject is also given in Wigner's book (Ref. 36).

Irreducible Representations

If there are no restrictions to the dimension n of matrices, the number of representations is infinite for a given group. This is no longer true if we only consider the so-called "irreducible representations" discussed in the following paragraph.

Consider an n-dimensional representation $D(R)$ of the group. A transformation of the following kind can be applied to each of the n matrices:

$$D'(R) = T\,D(R)\,T^{-1} \tag{16}$$

T is the transformation matrix and has the same dimension as $D(R)$. If any transformation matrix T can be found so that the transformed matrix $D'(R)$ has the form

$$D'(R) = \begin{pmatrix} [D_1(R)]_m & \vdots & 0 \\ \cdots\cdots\cdots & \vdots & \cdots\cdots\cdots \\ 0 & \vdots & [D_2(R)]_{n-m} \end{pmatrix}_n \tag{17}$$

then $D(R)$ is said to be "reducible," otherwise $D(R)$ is "irreducible." $D_1(R)$ and $D_2(R)$ are also representations of the group. Their matrices have the dimensions m and $n-m$, respectively, where $m < n$. If $D_1(R)$ and $D_2(R)$ cannot be further reduced by a transformation similar to (16), we say that $D(R)$ has been reduced to the irreducible representations $D_1(R)$ and $D_2(R)$. In the most general case an n-dimensional reducible representation can be reduced to the following form:

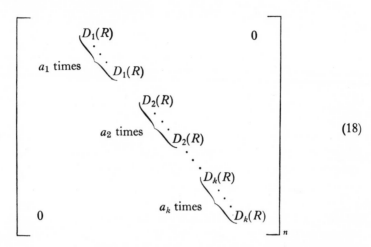

$$(18)$$

$D_1(R)$, $D_2(R)$,..., etc., are irreducible representations of the group, a_1 gives the number of times $D_1(R)$ appears in (18), and k is the number of different irreducible representations present in (18). The dimension of a representation $D_1(R)$ is always smaller than n. It can be as small as 1.

Orthogonality Relation

In this and the next subsection we will determine the number of nonequivalent irreducible representations for a group. It turns out to be always equal to or smaller than the order h of the group. We will show that it is equal to the number of classes. For Abelian groups each element forms a class by itself so that the number of classes and number of irreducible representations are equal to h.

First we will discuss the orthogonality relation. Let us consider two irreducible representations of the group E, A_2, A_3,..., R,..., A_h:

$$D^{(j)}(E), \quad D^{(j)}(A_2),...,D^{(j)}(R),...,D^{(j)}(A_h)$$
$$D^{(j')}(E), \quad D^{(j')}(A_2),...,D^{(j')}(R),...,D^{(j')}(A_h)$$
$$(19)$$

The orthogonality relation for the coefficients of the individual matrices can be expressed in the following form (Ref. 36):

$$\sum_R D^{(j)}(R)_{\nu\mu} D^{(j')}(R)^*_{\nu'\mu'} = \frac{h}{\sqrt{l_j l_{j'}}} \delta_{jj'} \delta_{\mu\mu'} \delta_{\nu\nu'}$$
$$(20)$$

where j and j' refer to two different representations and l_j is the dimension of the matrix $D^{(j)}(R)$. The summation is carried out over all group elements. The meaning of this equation can be visualized by considering $D^{(j)}(R)_{\nu\mu}$ as a component of a vector in an h-dimensional space. The h components are

$$D^{(j)}(E)_{\nu\mu}, \quad D^{(j)}(A_2)_{\nu\mu}, \ldots, D^{(j)}(R)_{\nu\mu}, \ldots, D^{(j)}(_hA)_{\nu\mu}$$

According to Eq. (20) all the possible scalar products of two different vectors are zero. This means that all the vectors are orthogonal to each other in the h-dimensional space. This puts a limitation to the dimension of the irreducible representation matrices as shown by the following considerations. The number of vectors for a given representation j is l_j^2 which is the number of coefficients in a matrix. To obtain the total number of vectors from all the irreducible representations we have to carry out a summation over all the representations j. This total number of vectors is equal to h since in an h-dimensional space only h vectors can be orthogonal. This fact is expressed by the following equation:

$$l_1^2 + l_2^2 + \cdots + l_c^2 = h \tag{21}$$

where c is the number of nonequivalent irreducible representations. For an Abelian group, where c is equal to h, all the values l_j are euqal to 1 which means that all irreducible representations are one-dimensional.

The Character of a Representation

The "spur" $\chi^{(j)}(R)$ of an l_j-dimensional matrix is the sum of the diagonal elements

$$\chi^{(j)}(R) = \sum_{\mu=1}^{l_j} D^{(j)}(R)_{\mu\mu} \tag{22}$$

The set of h numbers consisting of the spurs of all the matrices for a given representation j is called the "character" $\Gamma^{(j)}$ of the representation

$$\Gamma^{(j)} = \chi^{(j)}(E), \quad \chi^{(j)}(A_2), \ldots, \chi^{(j)}(R), \ldots, \chi^{(j)}(A_h) \tag{23}$$

The spur of a matrix is invariant to a transformation, which specifically means that conjugate elements, or elements belonging to the same class, have the same spur. Therefore, the character is already determined by k instead of h numbers, where k is the number of classes of the group:

$$\Gamma^{(j)} = \chi^{(j)}(C_1), \quad \chi^{(j)}(C_2), \ldots, \chi^{(j)}(C_\sigma), \ldots, \chi^{(j)}(C_k) \tag{24}$$

C_σ can be any element in the σth class.

We now will derive an orthogonality relation for the characters of two representations. We return to Eq. (20) and consider only diagonal elements of the matrices since all the products involving off-diagonal elements are zero. Equation (20) then becomes

$$\sum_R D^{(j)}(R)_{\nu\nu} D^{(j')}(R)_{\nu'\nu'}^* = \frac{h}{\sqrt{l_j l_{j'}}} \delta_{jj'} \delta_{\nu\nu'} \tag{25}$$

The summation over ν and ν' according to (22) leads to the following equation:

$$\sum_R \chi^{(j)}(R)\chi^{(j')}(R)^* = \frac{h}{\sqrt{l_j l_{j'}}}\delta_{jj'}\sum_{\nu=1}^{l_j}\sum_{\nu'=1}^{l_{j'}}\delta_{\nu\nu'} = h\,\delta_{jj'} \qquad (26)$$

From the comments to (23) and (24) we know that elements in the same class have the same character. Therefore, Eq. (26) can be written in the form

$$\sum_{\sigma=1}^{k}\chi^{(j)}(C_\sigma)\chi^{(j')}(C_\sigma)^* g_\sigma = h\,\delta_{jj'} \qquad (27)$$

where g_σ is the number of elements in the class C_σ. Equation (27) is an orthogonality relation for the character $\Gamma^{(j)}$ of the jth irreducible representation. This can be easily seen if $\Gamma^{(j)}$ in the form of Eq. (24) is considered as a vector in a k-dimensional space. Since, in such a space, only k vectors can be orthogonal, the number of characters and, therefore, the number of irreducible representations is equal to k. This fact is expressed by the basic theorem of representation theory:

> The number of nonequivalent irreducible representations c is equal to the number of classes k of the group.

The dimension l_j of each of the k (or c) irreducible representations is uniquely determined by (21), which means that there is only one way the k numbers l_j can be chosen so that the sum of their square is equal to h.

The actual form and especially the characters of the k irreducible representations can be obtained from equations given by Bethe (Ref. 1) and summarized by Rosenthal and Murphy (Ref. 27) for symmetry groups. Character tables can be found in many textbooks of spectroscopy (see, e.g., Ref. 11).

C. SYMMETRY GROUPS

The importance of group theory for spectroscopy lies in the fact that all molecules belong to certain symmetry groups. The symmetry of a molecule in its equilibrium position is determined by a set of symmetry elements, which are the elements of a symmetry group. The symmetry of nonpolymeric free molecules can be characterized by point groups while molecular as well as ionic crystals have space group symmetry. The symmetry elements for a chain molecule form a one-dimensional space group, which is sometimes called a "line group" (Ref. 35). In this subsection we will discuss various symmetry groups with special emphasis on the line groups.

Point Groups

Symmetry operations of a point group, when applied to a molecule, leave at least one point in the molecule invariant. Point groups have been discussed extensively by various authors over the past thirty years. A detailed and complete account of this subject is given in Herzberg's book (Ref. 11).

In this section we will discuss only one example of a point group as an illustration of how the symmetry of a molecule can be characterized by a group. We have chosen the planar symmetric trichlorobenzene molecule as shown in Fig. 1. The symmetry group is D_{3h}. It consists of the following 12 symmetry operations: the unity operation E, a reflection σ_h at the plane of the molecule, three reflections σ_v at the three planes of symmetry perpendicular to the plane of the molecule, three rotations C_2 around the twofold axes, two rotations C_3^1 and C_3^2 around a threefold

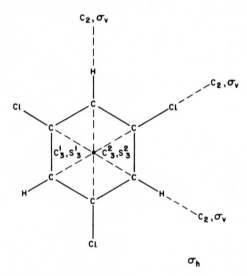

FIG. 1. 1,3,5-Trichlorobenzene molecule; point group D_{3h}.

rotation axis perpendicular to the plane of the molecule, and 2 threefold rotation-reflections S_3^1 and S_3^2 around the same axis. C_3^1 and S_3^1 correspond to rotations of 120°, while C_3^2 and S_3^2 are rotations of 240° or $-120°$. The 12 elements of the group can be divided into six classes as shown in Table I (the number of symmetry operations in each class is given in parentheses). The group has, therefore, six irreducible representations, four of them are one-dimensional and two are two-dimensional since Eq. (21) can only be satisfied in the following way:

$$1^2 + 1^2 + 1^2 + 1^2 + 2^2 + 2^2 = 12 \tag{28}$$

The characters of the six representations are also given in Table I. The right-hand side of the table will be explained later when we shall discuss the number and spectral activity of the normal vibrations belonging to each irreducible representation.

TABLE I

CHARACTER TABLE FOR POINT GROUP D_{3h}; SYMMETRIC TRICHLOROBENZENE MOLECULE

Characters			Six classes						a_i	Normal vibrations		Spectral activity	
$\Gamma^{(i)}$	Conventional Nomenclature	l_j	(1)E	(2)C_3	(3)C_2	(1)σ_h	(2)S_3	(3)σ_v		Translations	Rotations	IR	Raman
$\Gamma^{(1)}$	A_1'	1	+1	+1	+1	+1	+1	+1	4	0	0		$\alpha_{xx},\ \alpha_{yy},\ \alpha_{zz}$
$\Gamma^{(2)}$	A_1''	1	+1	+1	+1	−1	−1	−1	0	0	0		
$\Gamma^{(3)}$	A_2'	1	+1	+1	−1	+1	+1	−1	4	0	1 R_z		
$\Gamma^{(4)}$	A_2''	1	+1	+1	−1	−1	−1	+1	4	1 T_z	0	M_z	
$\Gamma^{(5)}$	E'	2	+2	−1	0	+2	−1	0	8(16)^a	1(2)$^a\ T_x,\ T_y$	0	$M_x,\ M_y$	$\alpha_{xx},\ \alpha_{yy},\ \alpha_{xy}$
$\Gamma^{(6)}$	E''	2	+2	−1	0	−2	+1	0	4(8)^a	0	1(2)$^a\ R_x,\ R_y$		$\alpha_{xz},\ \alpha_{yz}$

^{a} The numbers in parentheses indicate that doubly degenerate vibrations are counted twofold.

The Lattice Translation Group

Consider the three-dimensional lattice of a crystal or the one-dimensional lattice of a linear chain molecule with well-defined repeat units. A hypothetical lattice that extends to infinity is invariant under a translation by any multiple or combination of the three fundamental lattice vectors. Such translations can, therefore, be considered symmetry operations for the lattice. The composite of all these symmetry operations which leave the lattice invariant form a group of infinite order, called the "lattice translation group." In general, an element of this group is a translation $\mathbf{T}_{n_1 n_2 n_3}$ which corresponds to a vector of the form

$$\mathbf{T}_{n_1 n_2 n_3} = n_1 t_1 + n_2 t_2 + n_3 t_3 \tag{29}$$

where t_1, t_2, t_3 are the fundamental lattice vectors in a three-dimensional lattice and n_1, n_2, n_3 are integers. The product of two group elements corresponds to a simple addition of the vectors

$$\mathbf{T}_{n_1 n_2 n_3} \times \mathbf{T}_{l_1 l_2 l_3} = \mathbf{T}_{n_1 + l_1, n_2 + l_2, n_3 + l_3} \tag{30}$$

and the unit element is $\mathbf{T}_{000}$.

A real crystal or polymer chain is always finite, and a translation is no longer a "perfect" symmetry operation since the unit cells at one edge of a crystal will not be replaced by others upon a translation $\mathbf{T}$. It is possible to get around this difficulty by arbitrarily introducing Born's cyclic boundary conditions (see, e.g., Ref. 4), in which case a finite crystal is considered where a translation $\mathbf{T}_{N_1 N_2 N_3}$ is equivalent to the unit element $\mathbf{T}_{000}$ of the translation group. Furthermore,

$$\mathbf{T}_{N_1 N_2 N_3} = \mathbf{T}_{00 N_3} = \mathbf{T}_{0 N_2 0} = \mathbf{T}_{N_1 00} = \mathbf{T}_{000} \tag{31}$$

The order of the translation group is now finite and equal to $N_1 N_2 N_3$. The boundary conditions can also be interpreted by assuming an infinite fictitious crystal with periodic repeat units consisting of $N_1 N_2 N_3$ unit cells. For the one-dimensional case this condition can easily be visualized in a somewhat different way. In a one-dimensional finite lattice the order of the translation group is N and the fundamental lattice vector is t as shown in Fig. 2a. To fulfill the cyclic boundary condition the long, straight chain is bent to form a closed circle as shown in Fig. 2b. Each translation $\mathbf{T}_n$ (where $n = 1, 2, \ldots, N$) is now a true symmetry operation for the chain and Eq. (31) is trivial for this one-dimensional case. It simply becomes

$$\mathbf{T}_N = \mathbf{T}_0 \tag{31a}$$

It is important to realize that the cyclic boundary condition was invented by Born for mathematical convenience.

Before discussing the irreducible representations of the translation group we introduce a vector set reciprocal to the fundamental lattice vectors t_1, t_2, t_3. This reciprocal vector b_1, b_2, b_3 is defined by the following relation:

$$b_i t_k = \delta_{ik} \qquad \text{where} \qquad i, k = 1, 2, 3 \tag{32}$$

For a one-dimensional lattice Eq. (32) reduces to

$$|\mathbf{b}| = \frac{1}{|\mathbf{t}|} \tag{32a}$$

and $\mathbf{b}$ is a vector having the same direction as $\mathbf{t}$.

We now will proceed to discuss the irreducible representations of the translation group. First we note that the group is Abelian since the product of any two elements is a summation of two vectors, and such a summation is commutative. In an Aebelian group each element forms a class by itself so that the number of irreducible representations (number of classes) is equal to the order of the group, which is $N_1 N_2 N_3$. From Eq. (21) it follows that all the representation matrices

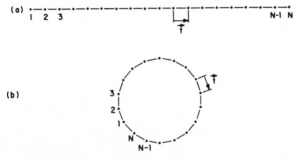

FIG. 2. (a) One-dimensional lattice of N repeat units. (b) Same as (a) but folded to a circle to demonstrate the cyclic boundary condition. t is the fundamental lattice vector.

are one-dimensional ($l_j = 1$, $c = h$) and, therefore, are simply numbers. To determine these numbers, we consider the jth representation with the character $\Gamma^{(j)}$, where

$$\Gamma^{(j)} = \chi^{(j)}_{100}, \quad \chi^{(j)}_{200}, \cdots, \chi^{(j)}_{N_100}, \cdots, \chi^{(j)}_{N_1 N_2 N_3} \tag{33}$$

The elements of the representation have to obey the same multiplication rules as the elements of the group itself [Eq. (30)]:

$$\chi^{(j)}_{n_1 n_2 n_3} \times \chi^{(j)}_{l_1 l_2 l_3} = \chi^{(j)}_{n_1 + l_1, \, n_2 + l_2, \, n_3 + l_3} \tag{34}$$

Furthermore, the spur of the unit element is 1 by definition:

$$\chi^{(j)}_{000} = \chi^{(j)}_{N_100} = \chi^{(j)}_{0N_20} = \cdots = \chi^{(j)}_{N_1 N_2 N_3} = 1 \tag{35}$$

It turns out that these two conditions are sufficient to determine the general form of the representation. Only an exponential function can fulfill Eqs. (34) and (35) simultaneously. Such a function has the following form (with the index j substituted by $\varkappa$):

$$\chi^{(p_1, p_2, p_3)}_{n_1, n_2, n_3} = \chi^{(\varkappa)}_{\mathbf{T}} = \exp[2\pi i(\varkappa \cdot \mathbf{T})] \tag{36}$$

where **T** is the translation vector given by Eq. (29) and **ᴋ** is the so-called "wave-number vector" defined by

$$\mathbf{ᴋ} = \frac{p_1}{N_1} b_1 + \frac{p_2}{N_2} b_2 + \frac{p_3}{N_3} b_3 \tag{37}$$

p_1, p_2, p_3 are integers running from 0 to $N_1 - 1$, $N_2 - 1$, $N_3 - 1$, respectively, and b_1, b_2, b_3 are the reciprocal lattice vectors defined by Eq. (32). Equation (36) represents a set of $(N_1 N_2 N_3)^2$ numbers. For a one-dimensional lattice this number reduces to N^2, and Eq. (36) becomes

$$\chi_n^{(p)} = \exp\left[2\pi i \left(\frac{p}{N} \mathbf{b} \cdot n\mathbf{t}\right)\right] = \exp\left(2\pi i \frac{pn}{N}\right) \tag{36a}$$

The character table for this case is given explicitly in Table II. The right-hand side of the table, concerning the number of normal vibrations and their spectral activity, will be discussed in Sections 3B and 4E.

Space Groups

So far we have considered only the translational symmetry of a lattice. Many lattices have additional symmetry elements, R, such as rotations, reflections, inversions, screw-rotations, and glide-reflections. Let us assume that a lattice contains H different symmetry elements of this kind (including the identity E). The total symmetry of the lattice is then described by a "space group," the symmetry operations of which are combinations of pure lattice translations with

TABL♦

CHARACTER TABLE FOR TH

Symmetr♦

$\Gamma^{(l)}$	$T_0 = E$	T_1	T_2
$\Gamma^{(0)}$	1	1	1
$\Gamma^{(1)}$	1	$\exp\left(2\pi i ᴋ_1 T_1\right) = \exp\left(2\pi i \frac{1}{N}\right)$	$\exp\left(2\pi i ᴋ_1 T_2\right) = \exp\left(2\pi i \frac{2}{N}\right)$
$\Gamma^{(2)}$	1	$\exp\left(2\pi i ᴋ_2 T_1\right) = \exp\left(2\pi i \frac{2}{N}\right)$	$\exp\left(2\pi i ᴋ_2 T_2\right) = \exp\left(2\pi i \frac{4}{N}\right)$
$\vdots$	$\vdots$	$\vdots$	$\vdots$
$\Gamma^{(N-1)}$	1	$\exp\left(2\pi i ᴋ_{N-1} T_1\right) = \exp\left(2\pi i \frac{N-1}{N}\right)$	$\exp\left(2\pi i ᴋ_{N-1} T_2\right) = \exp\left(2\pi i \frac{2(N-1)}{N}\right)$

[a] The first number refers to a finite chain, while the number in parentheses corresponds to a♦

[b] a = active, i = inactive.

the H other symmetry elements. There are $N_1 N_2 N_3 H$ such combinations possible for a finite space group of a lattice fulfilling Born's boundary conditions. The order of this space group is, therefore, $N_1 N_2 N_3 H$, while the $N_1 N_2 N_3$ translations form a self-conjugate subgroup of the space group. This statement is equivalent to saying that any element of the translation group, for example, $\mathbf{T}_1$, conjugated with an element S of the space group is again a pure translation $\mathbf{T}_2$ as expressed by

$$S\mathbf{T}_1 S^{-1} = \mathbf{T}_2 \qquad (38)$$

We want to point out that the multiplication of space group elements is in general not commutative; otherwise $\mathbf{T}_1$ would be equal to $\mathbf{T}_2$.

The translation group consists of the elements

$$\mathbf{T}_{001}, \quad \mathbf{T}_{002}, \ldots, \mathbf{T}_{N_1 N_2 N_3} \qquad (39)$$

Its order is $N_1 N_2 N_3$ and its index with respect to the space groups is H according to Eq. (5). The translation group has, therefore, H nonequivalent cosets which form the factor group of the space group (see Section 2A). The H cosets can be explicitly written down in the following form:

$$
\begin{aligned}
F_1 &= E\mathbf{T}_{001}, & E\mathbf{T}_{002}, \ldots, & \ E\mathbf{T}_{N_1 N_2 N_3} \\
F_2 &= R_2\mathbf{T}_{001}, & R_2\mathbf{T}_{002}, \ldots, & \ R_2\mathbf{T}_{N_1 N_2 N_3} \\
&\vdots & \vdots \quad \vdots & \quad \vdots \\
F_H &= R_H\mathbf{T}_{001}, & R_H\mathbf{T}_{002}, \ldots, & R_H\mathbf{T}_{N_1 N_2 N_3}
\end{aligned}
\qquad (40)
$$

NE-DIMENSIONAL TRANSLATION GROUP

erations		Normal vibrations			Spectral activity[b]	
	T_{N-1}	a_l	a_l^{trans}	a_l^{rot}	IR	Raman
.	1	$3m$	3	$3(1)^a$	a	a
. .	$\exp(2\pi i\varkappa_1 T_{N-1}) = \exp\left(2\pi i\dfrac{N-1}{N}\right)$	$3m$	0	0	i	i
.	$\exp(2\pi i\varkappa_2 T_{N-1}) = \exp\left(2\pi i\dfrac{2(N-1)}{N}\right)$	$3m$	0	0	i	i
	$\vdots$					
.	$\exp(2\pi i\varkappa_{N-1} T_{N-1}) = \exp\left(2\pi i\dfrac{(N-1)^2}{N}\right)$	$3m$	0	0	i	i

inscriptly extended lattice.

Each line contains one coset consisting of $N_1 N_2 N_3$ space group elements. We will see in Section 3B that this is a very useful way of arranging the $N_1 N_2 N_3 H$ elements. The factor group is given by the H elements F_1, F_2, ..., F_H or by the abbreviated form of Eq. (40):

$$ET, \quad R_2 T, ..., R_H T \qquad (41)$$

It is conventional to go even one step further and call

$$E, \quad R_2, ..., R_H \qquad (42)$$

the factor group of the space group. Since the factor group is independent of the order of the translation group, the finite and infinite space groups have the same factor group. The factor group of a space group is always isomorphous with a point group and has, therefore, the same character table. The space group is the direct product of the factor group in the form of (42) and the pure translation group. Each space group element is the product of two individual elements of the factor group and the translation group, respectively. As pointed out earlier, such a multiplication is not commutative. Therefore, the number of classes (equal to the number of irreducible representations) of the space group is not necessarily the product of the number of classes of the translation and factor group; and an irreducible representation of the space group cannot simply be obtained by multiplying a translation group representation with a factor group representation.

At this point we will confine ourselves to discussing only those space group representations which are derived from factor group representations, since it turns out (see Section 4) that those are the only ones that contain potentially infrared-active or Raman-active fundamental vibrations. The space group representations derived from the factor group's representation are obtained by assigning each element of a coset F_i [Eq. (40)] the *same* matrix, that is, the matrix which corresponds to the element R_i in an irreducible representation of the factor group. There is another way of looking at this problem. A theorem of group theory says that the matrices of a group representation corresponding to subgroup elements always form a representation (not necessarily irreducible) of the subgroup. Since the translation group is a subgroup of the space group we can apply this theorem to the space group representations derived from the factor group. All the translation group representations formed in this fashion are identical and equal to the totally symmetric representation $\Gamma^{(0)}$ of Table II. This representation corresponds to a value of $\varkappa = 0$. The problem just discussed should become somewhat clearer when one-dimensional examples are considered.

Line Groups

Chain molecules consisting of a linear arrangement of repeat units can be characterized by a one-dimensional space group which is sometimes called a "line group" (Ref. 35). Such a characterization is only correct if a molecule can be considered a straight individual chain without any interaction with its neighbors. But if several chains are arranged parallel in a crystalline region of a polymer, there exists a vibrational interaction between adjacent chains. It is usually small compared with the interaction between adjacent repeat units within the same chain. A line group analysis can, therefore, be considered a first approximation while a rigorous treatment would involve the three-dimensional space group of the crystal lattice. We will discuss this problem further in Section 4. At this point

TABLE III

Line Groups Characterized by Their Factor Groups

Factor group order	Symmetry elements of factor group	Isomorphous point group
	(a) POINT SYMMETRY OPERATIONS ONLY	
1	E	C_1
2	E, i	C_i
2	E, σ_h	C_s
2	$E, \sigma_v'; E, \sigma_v''$	C_s
2	$E, C_2'; E, C_2''$	C_2
p	$E, (p-1)C_p$	C_p
$2p$	$E, (p-1)C_p, p\sigma_v$	C_{pv}
4	$E, \sigma_h, \sigma_v', C_2''; E, \sigma_h, \sigma_v'', C_2'$	C_{2v}
4	$E, \sigma_v', C_2', i; E, \sigma_v'', C_2'', i$	C_{2h}
4	E, σ_h, C_2, i	C_{2h}
4	E, C_2, C_2', C_2''	$D_2 \equiv V$
8	$E, \sigma_h, \sigma_v', \sigma_v'', C_2, C_2', C_2'', i$	$D_{2h} \equiv V_h$
	(b) POINT SYMMETRY OPERATIONS, SCREW ROTATIONS, AND GLIDE REFLECTIONS	
2	$E, \bar{\sigma}_v'; E, \bar{\sigma}_v''$	C_s
p	$E, (p-1)\bar{C}_p$	C_p
4	$E, \bar{\sigma}_v', \sigma_v'', \bar{C}_2; E, \sigma_v', \bar{\sigma}_v'', \bar{C}_2$	C_{2v}
4	$E, \bar{\sigma}_v', \bar{\sigma}_v'', C_2$	C_{2v}
4	$E, \sigma_h, \bar{\sigma}_v', C_2''; E, \sigma_h, \bar{\sigma}_v'', C_2'$	C_{2v}
4	$E, \bar{\sigma}_v', C_2', i; E, \bar{\sigma}_v'', C_2'', i$	C_{2h}
4	$E, \sigma_h, \bar{C}_2, i$	C_{2h}
4	$E, \bar{C}_2, C_2', C_2''$	$D_2 \equiv V$
8	$E, \sigma_h, \sigma_v', \bar{\sigma}_v'', \bar{C}_2, C_2', C_2'', i; E, \sigma_h, \bar{\sigma}_v', \sigma_v'', \bar{C}_2, C_2', C_2'', i$	$D_{2h} \equiv V_h$
8	$E, \sigma_h, \bar{\sigma}_v', \bar{\sigma}_v'', C_2, C_2', C_2'', i$	$D_{2h} \equiv V_h$

some of the more important line groups will be listed and as an example the line group for linear polyethylene and hypothetically planar poly(methylene oxide) will be discussed.

A list of the line groups is given in Table III after Bhagavantam and Venkatarayudu (Ref. 3). The factor group elements $E\mathbf{T}, R_2\mathbf{T}, \ldots R_H\mathbf{T}$ [see Eq. (41)] are sufficient to characterize a line group. For simplicity the operations $E, R_2, \ldots,$ R_H in Table III are called the factor group elements of the line group. The symbols in Table III are defined as follows (the chain axis coincides with the z-axis of a coordinate system):

E identity operation

σ_h reflection at the xy-plane

σ_v reflection at a plane parallel to the z-axis

σ_v' reflection at the yz-plane

σ_v'' reflection at the xz-plane

C_p rotation around the z-axis by the angle $k360°/p$, where $k = 1, 2, \ldots, p-1$

C_2' rotation around x-axis by $180°$

C_2'' rotation around y-axis by $180°$

i inversion at a center

$\bar{\sigma}_v'$ glide reflection at the yz-plane along the z-axis

$\bar{\sigma}_v''$ glide reflection at the xz-plane along the z-axis

$\bar{C}_p$ screw rotation around the z-axis by the angle $k360°/p$, where $k = 1,$ $2, \ldots, p-1$

The symbols for the point groups in the last column of Table III are the same as Herzberg's (Ref. 11).

We now will discuss two examples.

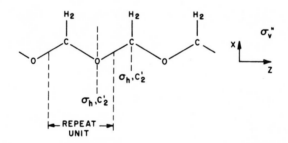

Fig. 3. Planar model of poly(methylene oxide) chain line group C_{2v}.

Hypothetically planar poly(methylene oxide): If we suppose the backbone of this very simple polymer to be a planar zig-zag chain† as shown in Fig. 3, the factor group is of the order 4. The symmetry operations are E, σ_h, σ_v'', C_2'. The point group isomorphous with the factor group is C_{2v}. The actual line group operations are given by the four cosets of the translation group:

$$
\begin{aligned}
T_1, \quad T_2, \quad &\ldots, T_N = E \\
\sigma_h T_1, \quad \sigma_h T_2, \quad &\ldots, \sigma_h T_N = \sigma_h \\
\sigma_v'' T_1, \quad \sigma_v'' T_2, \quad &\ldots, \sigma_v'' T_N = \sigma_v'' \\
C_2' T_1, \quad C_2' T_2, \quad &\ldots, C_2' T_N = C_2'
\end{aligned}
\tag{43}
$$

† In reality this chain is not planar. It is a helix with a crystallographic repeat unit of 17.3 Å containing nine chemical units —CH_2O— and five turns of the helix (Refs. 14, 34).

TABLE IV

CHARACTER TABLE FOR LINE GROUP C_{2v}; PLANAR POLYMETHYLENE OXIDE CHAIN

	Symmetry operations				Normal vibrations				Spectral activity	
	E	C_2'	σ_h	σ_v''	a_i	a_i^{skeleton}	a_i^{trans}	a_i^{rot}	IR	Raman
$\Gamma^{(1)} = A_1$	$+1$	$+1$	$+1$	$+1$	4	2	$1\,T_x$		M_x	$\alpha_{xx}, \alpha_{yy}, \alpha_{zz}$
$\Gamma^{(2)} = A_2$	$+1$	$+1$	-1	-1	1					α_{yz}
$\Gamma^{(3)} = B_1$	$+1$	-1	$+1$	-1	4	2	$1\,T_y$		M_y	α_{xy}
$\Gamma^{(4)} = B_2$	$+1$	-1	-1	$+1$	3	2	$1\,T_z$	$1\,R_z$	M_z	α_{xz}

TABLE V

CHARACTER TABLE FOR THE LINE GROUP V_h, POLYETHYLENE CHAIN

	Symmetry elements								Normal vibrations				Spectral activity	
	E	σ_h	$\bar{\sigma}_v'$	σ_v''	$\bar{C}_2$	C_2'	C_2''	i	a_i	a_i^{skeleton}	a_i^{trans}	a_i^{rot}	IR	Raman
$\Gamma^{(1)} = A_g$	$+1$	$+1$	$+1$	$+1$	$+1$	$+1$	$+1$	$+1$	3	1				$\alpha_{xx}, \alpha_{yy}, \alpha_{zz}$
$\Gamma^{(2)} = A_u$	$+1$	-1	-1	-1	$+1$	$+1$	$+1$	-1	1					
$\Gamma^{(3)} = B_{1g}$	$+1$	$+1$	-1	-1	$+1$	-1	-1	$+1$	3	1				α_{xy}
$\Gamma^{(4)} = B_{1u}$	$+1$	-1	$+1$	$+1$	$+1$	-1	-1	-1	2	1	$1\,T_z$		M_z	
$\Gamma^{(5)} = B_{2g}$	$+1$	-1	$+1$	-1	-1	-1	$+1$	$+1$	2	1		$1\,R_z$		α_{zz}
$\Gamma^{(6)} = B_{2u}$	$+1$	$+1$	-1	$+1$	-1	-1	$+1$	-1	3	1	$1\,T_y$		M_y	
$\Gamma^{(7)} = B_{3g}$	$+1$	-1	-1	$+1$	-1	$+1$	-1	$+1$	1					α_{yz}
$\Gamma^{(8)} = B_{3u}$	$+1$	$+1$	$+1$	-1	-1	$+1$	-1	-1	3	1	$1\,T_x$		M_x	

The character table for the factor group is shown in Table IV, the right-hand side of which will be discussed later.

Polyethylene: A planar zig-zag chain of polyethylene is shown in Fig. 4. The factor group of the line group has the order 8 and is isomorphous with the point group V_h. The character table is given in Table V after Tobin (Ref. 35). The number of normal vibrations and their spectral activity will be discussed in Sections 3 and 4.

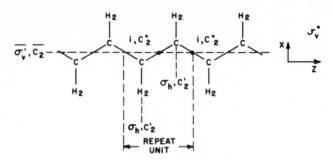

FIG. 4. Polyethylene chain; line group V_h.

3. Normal Modes of Vibration

A. Simple Polyatomic Molecules

The problem of describing the normal modes of vibration for polyatomic molecules has been treated by various authors. One of the early reviews of this subject was given by Rosenthal and Murphey (Ref. 27). In this subsection we will mainly follow their approach with the aim of applying it to chain molecules and crystals in Section 3B.

Cartesian and Normal Coordinates

We consider an individual N-atomic molecule with $3N$ normal vibrations (including translations and free rotations). To characterize these normal vibrations it is convenient to introduce two coordinate systems: Cartesian coordinates q_i and normal coordinates η_i, where $i = 1, 2, \ldots, 3N$, $q_1 = x_1$, $q_2 = y_1$, $q_3 = z_1$, $q_4 = x_2$, $q_5 = y_2$, $q_6 = z_2$, etc., x_1 being the deviation from the equilibrium position in the x-direction for atom number 1. The two coordinate systems are related by a linear transformation:

$$\eta_i = \sum_k c_{ik} q_k \qquad (44)$$

The numbers c_{ik} are the coefficients of a transformation matrix of the dimension $3N$. The potential energy V of the N-atomic molecule, expressed in Cartesian coordinates, is

$$V = \tfrac{1}{2} \sum_{i=1}^{3N} \sum_{j=1}^{3N} k_{ij} q_i q_j \qquad (45)$$

The normal coordinates η_i are defined by Eq. (44) where the values of c_{ik} are chosen so that if Eqs. (44) and (45) are combined, the potential energy V can be expressed in a simple quadratic form

$$V = \tfrac{1}{2} \sum_{i=1}^{3N} \lambda_i \eta_i^2 \tag{46}$$

It turns out that

$$\lambda_i = \omega_i^2 = (2\pi\nu_i)^2 \tag{47}$$

where ν_i is the frequency of the ith normal vibration. If the ith normal vibration is n_i-fold degenerate, that is, if n_i vibrations have the same frequency ν_i, Eq. (46) will assume the more general form

$$V = \tfrac{1}{2} \sum_{i=1}^{3N} \lambda_i(\eta_{i1}^2 + \eta_{i2}^2 + \cdots + \eta_{in_i}^2) \tag{48}$$

where the η_{i1}, η_{i2}, etc., are the n_i degenerate normal coordinates for the ith normal vibration.

A symmetry operation applied to a molecule leaves the potential energy V of a vibration invariant. This statement is the key to all the findings discussed in this section. V is only invariant if the sum in parenthesis in Eq. (48) stays invariant, which can be expressed by the following equation:

$$\eta_{i1}^2 + \eta_{i2}^2 + \cdots + \eta_{in_i}^2 = \eta_{i1}'^2 + \eta_{i2}'^2 + \cdots + \eta_{in_i}'^2 \tag{49}$$

where η_{i1}', η_{i2}', etc., are the normal coordinates after a symmetry operation R has been carried out. The meaning of Eq. (49) can be visualized in the following way. Consider the n_i normal coordinates of an n_i-fold degenerate vibration as the components of a vector in an n_i-dimensional space. Then Eq. (49) means that the length of the vector is invariant under a symmetry operation R. The most general linear transformation (it describes mathematically a symmetry operation R), applied to such a vector, has the form

$$\eta_{il}' = \sum_{k=1}^{n_i} C_i(R)_{lk} \eta_{ik} \tag{50}$$

$C_i(R)_{lk}$ is an element of the n_i-dimensional matrix $C_i(R)$. One such matrix corresponds to each symmetry element R of the group. The set of matrices

$$C_i(E), \quad C_i(A_2), \ldots, C_i(R), \ldots, C_i(A_h) \tag{51}$$

is, therefore, a representation of the symmetry group. It can be shown (Ref. 27) that the representation (51) is irreducible. Therefore, each normal vibration with the frequency ν_i corresponds to an irreducible representation C_i. Since a molecule has usually more vibrations (depending on the number of atoms) than there are irreducible representations for its symmetry group, several normal vibrations correspond to the same representation. Such vibrations are said to belong to the same symmetry type or to the same species.

Coordinate Transformations and Symmetry Operations

In the following paragraphs we will show how the number of normal vibrations belonging to each species can be calculated for any individual molecule. For this purpose we consider the transformation of the Cartesian coordinates under a symmetry operation R. The coordinates (deviations from the equilibrium position) of the N-atoms in the molecule are

$$x_1, \quad y_1, \quad z_1, \quad x_2, \quad y_2, \quad z_2, \quad \ldots, \quad x_N, \quad y_N, \quad z_N \tag{52a}$$

The symmetry operation transforms them into

$$x_1', \quad y_1', \quad z_1', \quad x_2', \quad y_2', \quad z_2', \quad \ldots, \quad x_N', \quad y_N', \quad z_N' \tag{52b}$$

The linear transformation corresponding to a symmetry operation R is of the form (Ref. 27)

$$
\begin{aligned}
x_1' &= g_{11}^{xx}(R)\,x_1 + g_{11}^{xy}(R)\,y_1 + g_{11}^{xz}(R)\,z_1 + g_{12}^{xx}(R)\,x_2 + \cdots \quad + g_{1N}^{xz}(R)\,z_N \\
y_1' &= g_{11}^{yx}(R)\,x_1 + g_{11}^{yy}(R)\,y_1 + \cdots \hspace{4.5cm} + g_{1N}^{yz}(R)\,z_N \\
z_1' &= \cdots \\
&\ \vdots \\
z_N' &= g_{N1}^{zx}(R)\,x_1 + g_{N1}^{zy}(R)\,y_1 + g_{N1}^{zz}(R)\,z_1 + g_{N2}^{zx}(R)\,x_2 + \cdots + g_{NN}^{zz}(R)\,z_N
\end{aligned}
\tag{53}
$$

Equation (53) can be written in the abbreviated form

$$\mathbf{q}' = G(R)\,\mathbf{q} \tag{54}$$

with $\mathbf{q}$ and $\mathbf{q}'$ being vectors with the components (52a) and (52b), respectively. $G(R)$ is a $3N$-dimensional transformation matrix associated with the operation R. The set of matrices

$$G(E), \quad G(A_2), \ldots, G(R), \ldots, G(A_h) \tag{55}$$

is a reducible representation of the symmetry group. By applying the proper similarity transformation [Eq. (16)], all the matrices $G(R)$ can be reduced to the form [see Eq. (18)]

$$
G(R) =
\begin{bmatrix}
\begin{array}{c} a_1 \text{ times} \begin{cases} C_1(R) \\ \quad \ddots \\ \quad\quad C_1(R) \end{cases} \end{array} & & & 0 \\
& a_2 \text{ times} \begin{cases} C_2(R) \\ \quad \ddots \\ \quad\quad C_2(R) \end{cases} & & \\
& & \ddots & \\
0 & & a_k \text{ times} & \begin{cases} C_k(R) \\ \quad \ddots \\ \quad\quad C_k(R) \end{cases}
\end{bmatrix}
\tag{56}
$$

where $C_1(R)$, $C_2(R)$, etc., are the k irreducible representations of the symmetry group, and a_i is the number of times $C_i(R)$ is contained in the representation $G(R)$. It turns out that a_i is also the number of normal vibrations associated with the irreducible representation $C_i(R)$ for the following reason: Equations analogous to (53) and (54) also hold if the Cartesian coordinates are replaced by normal coordinates. The transformation matrix can then also be reduced to the form of (56), and (53) reduces to $3N$ equations of the form of (50), where each normal coordinate corresponds to an irreducible representation $C_i(R)$. The number of equations containing $C_i(R)$ is a_i which is, therefore, the number of normal vibrations corresponding to $C_i(R)$. If a vibration is l_i-fold degenerate the dimension of $C_i(R)$ is also l_i and the a_i vibrations will have to be counted l_i-fold.

Number of Normal Vibrations Corresponding to an Irreducible Representation

We now will derive the important equation by which a_i can be computed. From Eq. (56) it is obvious that the spur $\chi(R)$ of the representation $G(R)$ is given by the following expression:

$$\chi(R) = \sum_{i=1}^{k} a_i \chi^{(i)}(R) \tag{57}$$

where $\chi^{(i)}(R)$ is the spur of the irreducible representation $C_i(R)$. We now multiply each side of Eq. (57) with $\chi^{(l)}(R)^*$ [= conjugate complex value of $\chi^{(l)}(R)$] and form the sum overall group elements R:

$$\sum_R \chi(R)\chi^{(l)}(R)^* = \sum_R \sum_{i=1}^{k} a_i \chi^{(i)}(R)\chi^{(l)}(R)^* \tag{58}$$

According to the orthogonality relation (26) Eq. (58) becomes

$$\frac{1}{h} \sum_R \chi(R)\chi^{(l)}(R)^* = a_l \tag{59}$$

This is the equation usually used to determine a_l, the number of normal vibrations corresponding to a certain irreducible representation. The spur $\chi^{(l)}(R)$ is obtained from the character table of the symmetry group. Only the evaluation of the spur $\chi(R)$ needs some further explanation.

Consider a symmetry operation R that moves atom l into the position of atom k. The new coordinates of this atom x'_k, y'_k, z'_k, will then be a function of the old ones x_l, y_l, z_l only, and the transformation (53) has the form

$$\begin{aligned}
&\vdots \\
x'_k &= 0+\cdots+0+g_{kl}^{xx}(R)\,x_l+g_{kl}^{xy}(R)\,y_l+g_{kl}^{xz}(R)\,z_l+0\cdots \\
y'_k &= 0+\cdots+0+g_{kl}^{yx}(R)\,x_l+g_{kl}^{yy}(R)\,y_l+g_{kl}^{yz}(R)\,z_l+0\cdots \tag{60} \\
z'_k &= 0+\cdots+0+g_{kl}^{zx}(R)\,x_l+g_{kl}^{zy}(R)\,y_l+g_{kl}^{zz}(R)\,z_l+0\cdots \\
&\vdots
\end{aligned}$$

The spur $\chi(R)$ of the transformation matrix $G(R)$ is the sum of the diagonal elements. Therefore, only those matrix elements for which $k = l$ will contribute to $\chi(R)$. In other words, the matrix elements will contribute to the spur only if atom k lies on the symmetry element and does not move during the symmetry operation R. As an example, we give the form of the transformation equation for the case where only one atom of a molecule, namely atom k, lies on the symmetry element and where the symmetry operation R is a rotation by the angle φ. Equation (60) then becomes

$$
\begin{aligned}
\vdots \\
x'_k &= 0 + \cdots + \cos\varphi\,x_k + \sin\varphi\,y_k + 0z_k + \cdots + 0 \\
y'_k &= 0 + \cdots - \sin\varphi\,x_k + \cos\varphi\,y_k + 0z_k + \cdots + 0 \\
z'_k &= 0 + \cdots + 0\quad x_k + \quad 0\quad y_k + z_k + \cdots + 0 \\
\vdots
\end{aligned}
\tag{61}
$$

The spur of this transformation is $2\cos\varphi + 1$. If the symmetry operation is a rotation-reflection, the spur becomes $2\cos\varphi - 1$. For the general case, where u atoms of a molecule lie on the symmetry element R, the spur of the transformation matrix becomes

$$
\chi(R) = u(2\cos\varphi \pm 1)
\tag{62}
$$

Where the plus $(+)$ sign holds for R being a pure rotation while the minus $(-)$ sign corresponds to a rotation-reflection. This includes a pure reflection and an inversion. Both are rotation-reflections where φ is equal to $0°$ and $180°$, respectively.

Equations (62) and (59) enable us to determine the number of normal vibrations a_l corresponding to an irreducible representation $\Gamma^{(l)}$. The pure translations and rotations are included in this evaluation and will have to be subtracted if we want to consider the $3N-6$ genuine normal vibrations only.† Equation (59) can also be used to evaluate the number of pure translations a_l^{trans} and rotations a_l^{rot} for each irreducible representation. $\chi(R)$ has to be replaced by the spur of the transformation of a pure translation, $\chi(R)^{\text{trans}}$, and a rotation, $\chi(R)^{\text{rot}}$, respectively. A translation can be represented by a vector. Under a symmetry operation R the three components, x, y, z, of a vector transform in the following way:

$$
\begin{aligned}
x' &= \quad \cos\varphi\,x + \sin\varphi\,y + 0z \\
y' &= -\sin\varphi\,x + \cos\varphi\,y + 0z \\
z' &= \quad 0\quad x + 0\quad y \pm z
\end{aligned}
\tag{63}
$$

The spur of this transformation is

$$
\chi(R)^{\text{trans}} = 2\cos\varphi \pm 1
\tag{64a}
$$

† It should be noticed that the pure rotations and translations are nongenuine normal vibrations (three of each) and not symmetry operations.

$\chi(R)^{\mathrm{rot}}$ can be obtained in a similar fashion by representing a rotation (nongenuine normal vibration) by its angular momentum. This will lead to

$$\chi(R)^{\mathrm{rot}} = \pm 2 \cos \varphi + 1 \qquad (64b)$$

The plus $(+)$ sign in Eqs. (64a) and (64b) hold if R is a rotation, the minus $(-)$ sign applies to a rotation-reflection as in the case of Eq. (62).

Example

As an example we consider again the symmetrical trichlorobenzene molecule, already discussed in Section 2C, and evaluate Eq. (59). The $\chi^{(l)}(R)$ values are the spurs given in Table I. From Eq. (62) we obtain the values of $\chi(R)$ and Eqs. (64a) and (64b) will lead to the corresponding values for pure translations and rotations.

TABLE VI

CHARACTERS AS COMPUTED BY EQS. (62), (64a), AND (64b) FOR SYMMETRIC TRICHLORO-BENZENE (FIG. 1). THE SYMMETRY GROUP IS D_{3h} WITH THE ORDER $h = 12$

Symmetry operations R	$u =$ Number of atoms on symmetry element	$\chi(R)$	$\chi(R)^{\mathrm{trans}}$	$\chi(R)^{\mathrm{rot}}$
E	12	36	3	3
$2C_3$	0	0	0	0
$3C_2$	4	-4	-1	-1
σ_h	12	12	1	-1
$2S_3$	0	0	-2	2
$3\sigma_v$	4	4	1	-1

The numerical values of these spurs are listed in Table VI. It now is possible to evaluate [with Eq. (59)] the number of genuine and nongenuine normal vibrations corresponding to each irreducible representation. The results are given in Table I.

B. CRYSTALLINE SYSTEMS

In condensed systems the molecular interaction will somewhat affect the absorption spectrum. This interaction is regular for crystalline systems. Therefore, it is possible to analyze its effect on the type and spectral activity of the normal vibrations. First we will discuss a crystal having the lowest space group symmetry.

Translation Group Symmetry

With Winston and Halford (Ref. 37) we consider a finite crystal consisting of $N_1 N_2 N_3$ unit cells. Each unit cell contains m atoms. The total number of normal vibrations of this crystal is, therefore, $3m N_1 N_2 N_3$. We now will derive the number of vibrations a_l corresponding to each irreducible representation of the symmetry group. The pure translation group has $N_1 N_2 N_3$ such representations as shown earlier. The number a_l can be determined by Eq. (59), where the order h of the group is $N_1 N_2 N_3$. The values of the spurs $\chi^{(l)}(R)$ are expressed by Eq. (36). R is now a translation $\mathbf{T}_{n_1 n_2 n_3}$. For the one-dimensional case the spurs are given in the character table, Table II. The transformation matrix, $G(R)$, given by Eq. (54), is $3mN_1 N_2 N_3$-dimensional. Its spur $\chi(R)$ can easily be evaluated (Ref. 37):

$$\chi(R) \begin{cases} = 3mN_1 N_2 N_3 & \text{if} \quad R = E \\ = 0 & \text{if} \quad R = \mathbf{T}_{n_1 n_2 n_3} \neq E \end{cases} \tag{65}$$

The first part of Eq. (65) is obvious since $G(E)$ is a $3mN_1 N_2 N_3$-dimensional unit matrix. If R is a pure translation, all the diagonal elements of $G(R)$ disappear, since none of the atoms is invariant under a translation. Therefore, the sum $\chi(R)$ of these diagonal elements also disappears. We now substitute the values of $\chi(R)$ from Eq. (65) into Eq. (59) and obtain the following simplified form:

$$a_l = \frac{1}{N_1 N_2 N_3} \chi(E) \chi^{(l)}(E)^* = \frac{1}{N_1 N_2 N_3} 3mN_1 N_2 N_3 = 3m \tag{66}$$

Since a_l is a constant (independent of l) every irreducible representation of the translation group contains $3m$ normal vibrations as indicated in Table II for the one-dimensional crystal. This enumeration includes the nongenuine "vibrations." They can be evaluated separately by using the $\chi(R)$ values given by Eqs. (64a) and (64b). If the symmetry operation R is a translation, φ becomes 0 and Eqs. (64) become ($+$ sign)

$$\chi(R)^{\text{trans}} = \chi(R)^{\text{rot}} = 3 \tag{67}$$

Evaluating Eq. (59) for a one-dimensional space group for which the $\chi^{(l)}(R)$ values are listed in Table II we obtain

$$a_l^{\text{trans}} = a_l^{\text{rot}} = \frac{1}{N} \sum_R 3\chi^{(l)}(R) = \begin{cases} 3 & \text{if} \quad l = 0 \\ 0 & \text{if} \quad l \neq 0 \end{cases} \tag{68}$$

a_l^{trans} and a_l^{rot} are the numbers of translations and rotations, respectively (both are nongenuine normal vibrations), corresponding to the lth irreducible representation. It is important to note that all the nongenuine normal vibrations belong to the totally symmetric representation $\Gamma^{(0)}$. Furthermore, we remark that only a finite crystal can have three free rotations while this would not be possible for an infinite lattice. A finite segment with the Born boundary conditions behaves like an infinite lattice as far as free rotations are concerned. A one dimensional infinite lattice has one free rotation as indicated in Table II.

Space Group Symmetry

In the last subsection we discussed a crystal with the lowest space group symmetry consisting of translational elements only. Now we will consider space groups containing other symmetry elements as well and try to determine the number of normal vibrations corresponding to each representation of the space group. This is a rather complex task since the space group representations are usually not known explicitly, although Winston and Halford (Ref. 37) have given expressions to calculate them as we will show in a later section.

If we now restrict ourselves to potentially infrared- or Raman-active vibrations we only will have to consider the representations of the space group derived from the factor group's representations. The number of normal vibrations correspond-ing to these space group representations can be calculated by a method also given by Winston and Halford (Ref. 37) which we now will describe.

The procedure is exactly the same as the one used earlier for the point and translation groups. We have to evaluate Eq. (59) which we rewrite for the space group in the form

$$a_l = \frac{1}{N_1 N_2 N_3 H} \sum_{R \text{ in sp. gr.}} \chi(R) \chi^{(l)}(R) \tag{69}$$

where R is now a space group element and $\chi^{(l)}(R)$ the spur of its corresponding matrix in the lth representation, and a_l is the number of normal vibrations corre-sponding to this representation. The spur of the transformation matrix $\chi(R)$ will be discussed below. It turns out to be convenient to split the summation into two steps in the following fashion:

$$a_l = \frac{1}{N_1 N_2 N_3 H} \sum_{\text{cosets}} \sum_{R \text{ in cosets}} \chi(R) \chi^{(l)}(R) \tag{70}$$

We now consider only the space group representations derived from the factor group's representation and note that the values of $\chi^{(l)}(R)$ are identical for space group elements of the same coset F_i [see Eq. (40) and discussion thereafter] and equal to the factor group representation $\chi^{(l)}(F_i)$. Therefore, Eq. (70) becomes

$$a_l = \frac{1}{N_1 N_2 N_3 H} \sum_{\text{cosets } F_i} \chi^{(l)}(F_i) \sum_{R \text{ in coset}} \chi(R) \tag{71}$$

It now remains to evaluate $\sum_{R \text{ in coset}} \chi(R)$. For this purpose, we consider the system of transformation equations [corresponding to Eq. (60). For a space group operation it has the dimension $3N_1 N_2 N_3 H$. Only the atoms left invariant in a symmetry operation R will contribute to diagonal elements in Eq. (61) and, therefore, to the spur $\chi(R)$ of the transformation matrix. The contribution to $\chi(R)$ of one such atom in one unit cell is $2 \cos \varphi \pm 1$, where φ is the angle of rotation corresponding to the symmetry operation R. The $\pm$ signs are explained after Eq. (62). The number of invariant atoms in one unit cell is called $u(\text{coset})$ (Ref. 37). It is determined by the specific structure of the crystal. As implied by the nomenclature, $u(\text{coset})$ is the same for any space group element of the same coset.

This will be clearer from the examples treated below. Since the crystal contains $N_1 N_2 N_3$ unit cells, we obtain

$$\sum_{R\,\text{in coset}} \chi(R) = N_1 N_2 N_3\, u(\text{coset})\,(2 \cos \varphi \pm 1) \qquad (72)$$

Equations (71) and (72) combined lead to the expression for the number of normal vibrations corresponding to the lth factor group representation:

$$a_l = \frac{1}{H} \sum_{\text{cosets } F_i} \chi^{(l)}(F_i)\, u(\text{coset})\,(2 \cos \varphi \pm 1) \qquad (73)$$

It is important to note that a_l does not depend on the number of unit cells considered, which means that the treatment of one unit cell only, as originally suggested by Bhagavantam and Venkatarayudu (Ref. 2), would lead to the same result. If we add the a_l values for all factor group representations we obtain $3m$, where m is the number of atoms per unit cell. This result is obvious if we reduce a space group to the pure translation group so that all the factor group representations will correspond to the totally symmetric representation $\Gamma^{(0)}$ of the translation group. The number of the vibrations that correspond to any translation group representation and, therefore, also to $\Gamma^{(0)}$ is $3m$ as shown in the previous subsection (see Table II).

Examples

Planar polymethylene oxide: The hypothetically planar zig-zag structure of this polymer is shown in Fig. 3, and the one-dimensional space group symmetry elements are listed in (43) arranged as the four cosets of the line group C_{2v}.

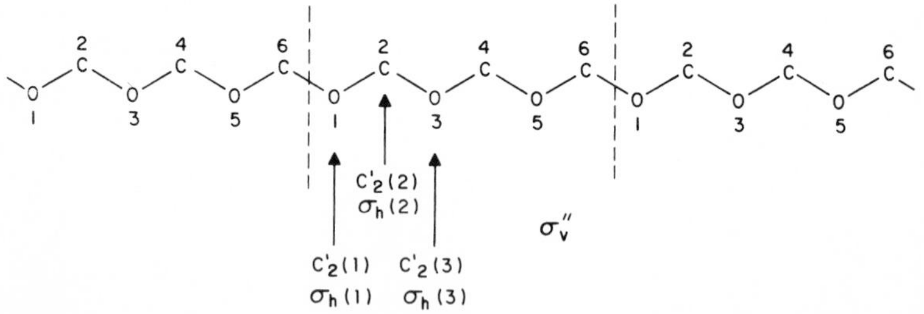

Fig. 5. Chain of hypothetically planar polymethylene oxide. The center section of three repeat units ($N = 3$) is arbitrarily chosen as the basic finite chain fulfilling the Born boundary conditions.

We will discuss this example in detail to clarify and illustrate the methods described in the preceding paragraphs. We arbitrarily consider a finite chain of three repeat units ($N = 3$) fulfilling the Born boundary conditions (see Section 2C). This is illustrated for the backbone chain of polymethylene oxide in Fig. 5, where the center section is considered as the basic finite chain. The pure translation

TABLE VII

Symmetry Operations for the E- and σ_h-Cosets of the Line Group C_{2v}

E-coset			σ_h-coset		
T_1	T_2	$T_3 = E$	$\sigma_h(1)$, $\sigma_h(2)T_2$, $\sigma_h(3)T_1$,	$\sigma_h(1)T_1$, $\sigma_h(2)$, $\sigma_h(3)T_2$,	$\sigma_h(1)T_2$, $\sigma_h(2)T_1$, $\sigma_h(3)$,
			$\sigma_h(4)$, $\sigma_h(5)T_2$, $\sigma_h(6)T_1$	$\sigma_h(4)T_1$, $\sigma_h(5)$, $\sigma_h(6)T_2$	$\sigma_h(4)T_2$, $\sigma_h(5)T_1$, $\sigma_h(6)$
1 ⟶ 3	1 ⟶ 5	1 ⟶ 1	1 ⟶ 1	1 ⟶ 3	1 ⟶ 5
2 ⟶ 4	2 ⟶ 6	2 ⟶ 2	2 ⟶ 6	2 ⟶ 2	2 ⟶ 4
3 ⟶ 5	3 ⟶ 1	3 ⟶ 3	3 ⟶ 5	3 ⟶ 1	3 ⟶ 3
4 ⟶ 6	4 ⟶ 2	4 ⟶ 4	4 ⟶ 4	4 ⟶ 6	4 ⟶ 2
5 ⟶ 1	5 ⟶ 3	5 ⟶ 5	5 ⟶ 3	5 ⟶ 5	5 ⟶ 1
6 ⟶ 2	6 ⟶ 4	6 ⟶ 6	6 ⟶ 2	6 ⟶ 4	6 ⟶ 6

[a] Shown is a specific example where the finite chain consists of three repeat units of the polymethyleneoxide skeleton. All the six symmetry operations at the head of each σ_h-column are identical. The numbers in the table indicate the positions of the six atoms before and after a symmetry operation.

group consists of three elements T_1, T_2, and T_3. The last one, which is a translation by three repeat units is equivalent with the identity operation E. The atoms in the finite chain segment are numbered accordingly. $C_2'(1)$ is the twofold axis through atom 1, $C_2'(2)$ is the same through atom 2, etc.; $\sigma_h(1)$ is a plane of symmetry through atom 1, while σ_v'' is a plane of symmetry defined by the plane of the zig-zag chain. The line group consists of combinations of these symmetry elements with T_1, T_2, and T_3. The number of such combinations is larger than the order of the line group; some of the combinations are, therefore, identical symmetry operations. The order of the line group is 12, and the elements expressed as the four cosets are given by (43), where $N = 3$. The identity of some of the elements is shown in Table VII. This table can easily be extended for the cosets of C_2' and σ_v''. It is clear that each element of the σ_h-coset has five equivalent elements.

TABLE VIII

			u(coset)	
Coset	φ (deg)	$2\cos\varphi \pm 1^a$	Backbone	Whole molecule
E	0	3	2	4
C_2'	180	-1	2	2
σ_h	0	1	2	4
σ_v''	0	1	2	2

^{a} $(+)$ sign for rotation; $(-)$ sign for rotation reflection.

We now will determine the number of normal vibrations a_l corresponding to each of the factor group representations (see Table IV) with the help of Eq. (73). The numerical values of $2\cos\varphi \pm 1$ and of u(coset) are listed in Table VIII for the backbone chain as well as for the whole molecule including the hydrogen atoms. u(coset) is the number of atoms left invariant in any one unit cell if all the symmetry operations (three) in a coset are applied to the chain. From Table VII it can easily be seen that this number is 2 for the E- and σ_h-coset if the backbone chain is considered only. The other u(coset) numbers listed in Table VIII can be obtained in a similar way. The a_l values are computed from Eq. (73). As an example one of the computations is given explicitly

$$a_1 = \tfrac{1}{4}[(+1)\times 4\times 3+(+1)\times 2\times(-1)+(+1)\times 4\times 1+(+1)\times 2\times 1] = 4 \quad (74)$$

The values for a_1, a_2, a_3, and a_4 are given in Table IV for the polymethylene oxide chain as well as for the skeleton alone.

Four of the normal vibrations are nongenuine: three translations and one rotation around the z-axis. The distribution of the translations among the repre-

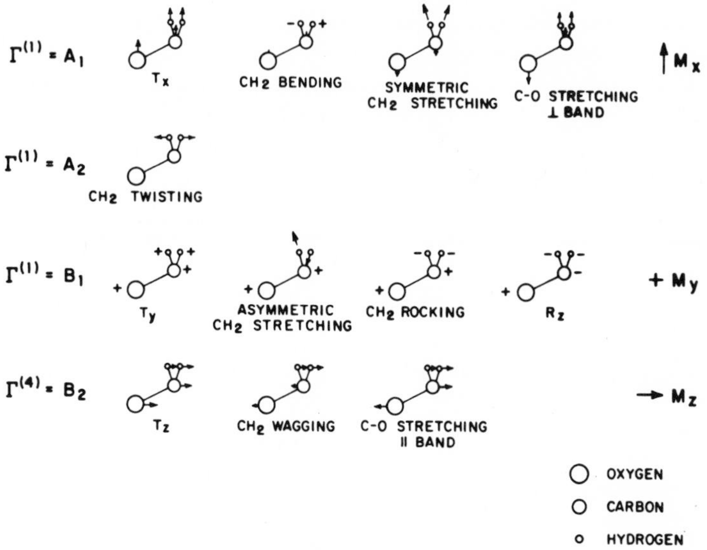

FIG. 6. Normal vibrations of the planar polymethylene oxide chain. Shown is one repeat unit. The CH_2 group lies in the xy-plane but is shown slightly twisted.

sentations $\Gamma^{(l)}$ can either be obtained formally with the help of Eq. (64a), as in the case of an ordinary polyatomic molecule, or by simply inspecting the character table (Table IV). The translation in the x-direction, for example, belongs to the representation $\Gamma^{(1)} (= A_1)$ since this translation is totally symmetric with respect to any of the four factor group symmetry elements. Rotations around the x- and

TABLE IX

Coset	φ (deg)	2 cos $\varphi \pm 1$[a]	u(coset) Backbone	u(coset) Whole molecule
E	0	3	2	6
σ_h	0	1	2	6
$\bar{\sigma}_v'$	0	1	0	0
σ_v''	0	1	2	2
$\overline{C}_2$	180	-1	0	0
C_2'	180	-1	2	2
C_2''	180	-1	0	0
i	180	-3	0	0

[a] See footnote to Table VIII.

y-axes are impossible since we deal essentially with an infinitely long chain (because of the Born boundary conditions applied to a finite segment). The rotation R_z around the z-axis is symmetrical with respect to elements of the cosets E and σ_h, but antisymmetrical with respect to C_2' and σ_v''. It, therefore, belongs to the representation $\Gamma^{(4)}$ $(= B_2)$. The geometrical form of the 12 factor group normal vibrations is shown for one unit cell in Fig. 6.

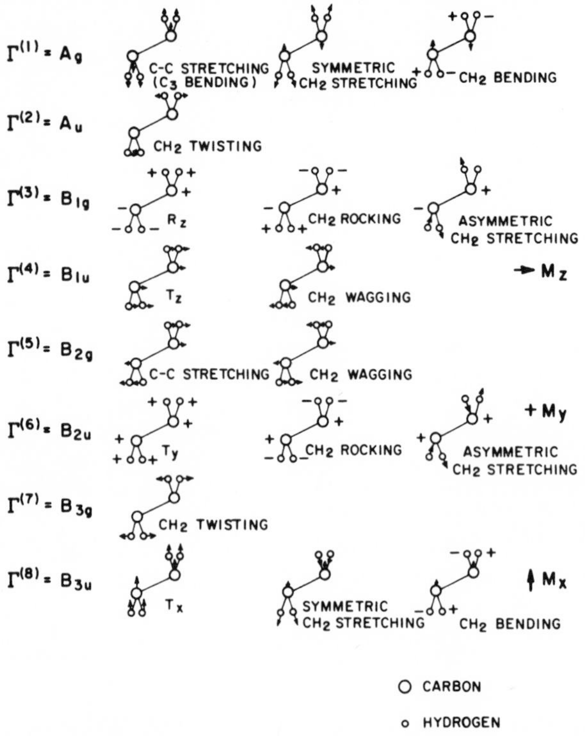

FIG. 7. Geometrical form of the 13 fundamental vibrations of the polyethylene chain. As in Fig. 6 the CH₂ groups are shown slightly twisted with respect to the xy-plane.

Polyethylene: This is the most thoroughly analyzed polymer as far as its infrared spectrum is concerned. The factor group and its representations are given as an example in Section 2C (Table V). A description of the normal vibrations is obtained in exactly the same way as for polymethylene oxide. In analogy to Table VIII the values for $2\cos\varphi \pm 1$ and $u(\text{coset})$ are given in Table IX for the line group V_h. The number of genuine and nongenuine normal vibrations a_l corresponding to the irreducible factor group representation $\Gamma^{(l)}$ can now be calculated as in the previous example. The results are given in Table V and the geometrical form of the vibrations is shown in Fig. 7.

4. Selection Rules

A. GENERAL SELECTION RULES FOR INFRARED AND RAMAN SPECTRA

Selection rules which determine whether molecular vibrations are infrared- or Raman-active are discussed in many textbooks on spectroscopy. We, therefore, will not give a detailed derivation of these rules but merely the results and show how they can be applied to crystalline and polymeric systems.

A normal vibration that is connected with a change in electric dipole moment is infrared-active. It is Raman-active if a change in electrical polarizability occurs during the vibration. These are the well-known classical selection rules.

In a quantum theoretical treatment these selection rules can be expressed in the following way (Ref. 11). Consider a transition between two vibrational levels v' and v''. The vibrational transition probability is proportional to the square of the transition moment $[M]^{v'v''}$ expressed by the space integral

$$[M]^{v'v''} = \int \psi_{v'} \psi_{v''}^* \mathbf{M} \, d\tau \tag{75}$$

where $\mathbf{M}$ is a dipole moment vector with the components M_x, M_y, M_z, and $\psi_{v'}$ and $\psi_{v''}$ are the eigenfunctions of the two vibrational levels. In a nonsymmetrical molecule all transitions are active. In a symmetric molecule only those transitions are infrared-active for which at least one component of the integrand

$$\psi_{v'} \psi_{v''}^* \mathbf{M} \tag{76}$$

is totally symmetrical, that is, remains unchanged if any one of the symmetry operations is applied to the molecule. The selection rules for the Raman effect are analogous: A transition is Raman-active if at least one of the six integrands

$$\psi_{v'} \psi_{v''}^* \alpha_{ik} \tag{77}$$

is totally symmetrical, where α_{ik} can be any of the six components α_{xx}, α_{yy}, α_{zz}, α_{xy}, α_{zy}, α_{yz} of the polarizability tensor.

B. SYMMETRY PROPERTIES OF VARIOUS ENTITIES

To facilitate the understanding of the following developments we inject this subsection with a discussion of the symmetry properties of various entites we are concerned with. Consider a molecule or a crystal characterized by a certain symmetry group. Such a group can further be characterized by a character table as shown for specific examples in Tables I, II, IV, and V. Every one of the entities $\psi_{v'}$, $\psi_{v''}$, M_x, M_y, M_z, α_{xx}, etc., or one of the normal vibrations corresponds to one of the group representations† and, therefore, has the symmetry behavior determined by the corresponding character. This can easily be visualized for one-dimensional representations of point groups where the value of the spur can only

† "Irreducible representation," "symmetry type," and "species" are names for one and the same thing.

be $+1$ or -1 meaning that the entity considered is either symmetrical or anti-symmetrical with respect to the corresponding symmetry element. As an example we consider the point group D_{3h}, the character table of which is given in Table I. The component M_z of the dipole moment is a vector in the z-direction. It, therefore, behaves like a translation T_z with respect to a symmetry operation. M_z belongs to the representation A_2'' ($= \Gamma^{(4)}$) since the direction of the vector does not change if any one of the symmetry operation E, C_3, and σ_v is applied, but it changes sign for the symmetry operations C_2, σ_h, and S_3. In the case of a two-dimensional representation ($=$ twofold degenerate symmetry type) the situation is not quite as simple. From Table I we see that the vectors M_x and M_y belong to the character E' ($= \Gamma^{(5)}$). This means that during a symmetry operation R the two vectors undergo a transformation of the following form:

$$\begin{aligned} M_x' &= M_x D^{(5)}(R)_{11} + M_y D^{(5)}(R)_{12} \\ M_y' &= M_x D^{(5)}(R)_{21} + M_y D^{(5)}(R)_{22} \end{aligned} \tag{78}$$

where M_x' and M_y' are the components of the dipole moment after the symmetry operation R. According to Eq. (22) the spur $\chi^{(5)}(R)$ is given by

$$\chi^{(5)}(R) = D^{(5)}(R)_{11} + D^{(5)}(R)_{22} \tag{79}$$

Its value for various symmetry operations is listed in Table I. Similar relations hold for groups with threefold degenerate symmetry types.

In Section 3 we showed that every normal vibration belongs to a certain representation or symmetry type. This symmetry type determines the behavior of the corresponding normal coordinate η with respect to a symmetry operation. Once this is known it is easy to find the symmetry type for the eigenfunction ψ_v, at least for nondegenerate vibrations. For a harmonic oscillator ψ_v is proportional to the Hermite polynomial of the vth degree, where v is the vibrational quantum number:

$$\psi_v \sim H_v(\eta) \tag{80}$$

where $H_0(\eta) = 1$, $H_1(\eta) = 2$, $H_2(\eta) = 4\eta^2 - 2$, $H_3(\eta) = 8\eta^3 - 12$, etc. From this it follows that ψ_0 (ground state), ψ_2 (doubly excited state), ψ_4, etc., are even functions in η and, therefore, belong to the totally symmetric representation. The eigenfunctions ψ_1 (first excited state), ψ_3, ψ_5, etc., are odd functions in η and, therefore, belong to the same representation as η itself.

If we know the symmetry type for two entities, say two eigenfunctions, the representation of their product is obtained by multiplying the respective entries n the character table. If, for example, the two eigenfunctions belong to the symmetry types A_1'', and A_2', respectively (Table I), their product belongs to A_2''. This rule holds only for nondegenerate symmetry types. In the case of degenerate ones, the situation is more complicated since the product of two entities may belong to more than just one symmetry type. Tables which facilitate these multiplications are given by Herzberg (Ref. 11, pp. 126–129).

The symmetry types for the components of the polarizability tensor α_{ik} may be obtained in a similar way (Ref. 11, p. 254) since α_{ik} is defined by

$$P_i = \alpha_{ik} E_k \tag{81}$$

E_k and P_i are the components of the electric vector of the radiation and the induced dipole moment, respectively. The symmetry types of E_k and P_i are known since they both are components of vectors. The product of the two symmetry types determines the symmetry type of α_{ik}. Complications arise again in the case of degenerate symmetry types as discussed in Ref. 11 (pp. 255–256). For specific examples the assignment of the six components α_{ik} to symmetry types is given in Tables I, II, IV, and V.

C. Specific Selection Rules

From Section 4A we know that transitions are only infrared- or Raman-active if any component of expression (76) or (77), respectively, belongs to the totally symmetrical representation. The symmetry types of these products can easily be obtained as discussed in the previous subsection.

First we consider fundamental vibrations where the transition takes place between the ground state ψ_0 and a first excited state ψ_1. ψ_0 is totally symmetrical and ψ_1 belongs to the same representation as the normal coordinate η or the vibration itself as shown in Section 4B. The product $\psi_0 \psi_1^*$ has, therefore, the same symmetry type as the vibration itself. If we multiply this symmetry type with the symmetry type of M_i or α_{ik} we obtain the symmetry type of expressions (76) or (77), respectively. Such a product of two symmetry types can only contain the totally symmetrical type if the two symmetry types are identical. This rule can easily be checked for nondegenerate symmetry types by inspection of the character tables, but it also holds for degenerate ones (see Ref. 11, pp. 126–129). The fundamental of a normal vibration can, therefore, only be infrared- or Raman-active if it belongs to the same symmetry type as a component of the dipole moment vector $\mathbf{M}$ or the polarizability tensor α_{ik}, respectively, since only then are the products (76) and (77) totally symmetrical.

The selection rules for overtone and combination bands are obtained in the same fashion. First we have to find the symmetry type of the vibrational eigenfunctions involved and then determine whether any one of the products (76) and (77) contains a totally symmetric species. The first overtone of a normal vibration is a transition between the states ψ_0 and ψ_2. For groups with no degenerate species both states are totally symmetric. Therefore, (76) or (77) are only totally symmetric (and the transition infrared- or Raman-active) if $\mathbf{M}$ or α_{ik} have a totally symmetric component themselves. The wave function ψ_3 has the same symmetry as ψ_1 (see Section 4B) and the activity of the second overtone is the same as for the fundamental vibration. For degenerate vibrations the symmetry of the higher vibrational levels often involves several species. A table given by Herzberg (Ref. 11, p. 127) facilitates the determination of those. For a combination band the upper state is a state where one vibration is n-fold and another one m-fold excited and has the wave function ψ_{n+m} the symmetry of which is the symmetry of the product $\psi_n \psi_m$. Knowing this fact selection rules are obtained as before. Finally we mention that a transition where $\psi_{v'}$ and $\psi_{v''}$ are two excited states can be treated in the same fashion. The selection rules discussed so far will now be applied to more specific systems.

D. Point Groups

For point groups the vibrational selection rules have been described extensively (see, e.g., Ref. 11). We, therefore, will not do so in any detail, but merely apply the rules to the specific example of the symmetric trichlorobenzene molecule as an illustration. The infrared- and Raman-activities for the fundamental vibrations are given in Table I. Four vibrations of species A_2'' and eight vibrations (doubly degenerate) of species E' are infrared-active while four vibrations of species A_1', eight vibrations of species E' and four vibrations of species E'' are Raman-active. The four vibrations of species A_2' are neither infrared- nor Raman-active.

We will now consider just a few examples of overtones and combinations. Any overtone of a vibration of the totally symmetric species A_1' belongs to the same species and is, therefore, only Raman-active. The first overtone of any one of the four completely inactive vibrations of species A_2' is totally symmetric and should, therefore, show up in the Raman effect, while the second overtone belongs to the same species as the fundamental and is, therefore, expected not to be infrared- or Raman-active. The first overtone of a degenerate infrared inactive vibration of species E'' belongs to the two species A_1' and E' and is, therefore, infrared- as well as Raman-active. Finally, we consider as an example a combination band of two singly excited vibrations of species A_1' and E'. This vibration is infrared- as well as Raman-active since it belongs to the species E'.

E. The Lattice Translation Group

The selection rules derived in this and the following subsection hold for one-dimensional as well as three-dimensional lattices.

First we summarize the contents of Sections 2C and 3B concerning the properties of the translation group. For a three-dimensional lattice the order of the group is $N_1 N_2 N_3$ which is the number of unit cells in an arbitrarily chosen crystalline block. The group is Abelian and, therefore, the number of irreducible representations is equal to the order of the group. These representations, characterized by a wave-number vector $\varkappa$, are one-dimensional and their characters are given by Eq. (36). The unit cell contains m atoms so that the total number of normal vibrations is $3mN_1 N_2 N_3$. They are equally divided among all the representations. The nongenuine normal vibrations such as pure translations belong to the totally symmetrical representation $\Gamma^{(0)}$, that is, the representation for which $\varkappa = 0$.

Actually, any vector or vector product is totally symmetric with respect to the translation group which means that all the components of the dipole moment **M** and the polarizability tensor α_{ik} belong to the representation $\Gamma^{(0)}$. For the one-dimensional translation group these facts are all summarized in Table II.

Selection Rules for Fundamentals

With the discussion of Sections 4A to 4C in mind we can immediately give the selection rules for the fundamental vibrations of a molecular crystal with translational symmetry. Only those transitions are infrared- or Raman-active for which expression (76) or (77) is totally symmetrical. Since the ground state ψ_0 and any component of **M** and α_{ik} belong to the representation $\Gamma^{(0)}$, the upper state ψ_1 has to belong to the same representation. Therefore, only the $3m$ vibrations that belong to $\Gamma^{(0)}$ are infrared- or Raman-active; all the other fundamental vibrations are inactive. The number of active fundamentals only depends upon m, the number of atoms per unit cell. It is particularly important to note that the spectral activity of fundamental vibrations does not depend on an arbitrarily chosen size of the crystal blocks containing $N_1 N_2 N_3$ unit cells as long as the Born boundary condition is fulfilled.

The selection rule has a very simple geometrical interpretation. If we say that a vibration is totally symmetrical with respect to a translation (symmetry operation of the translation group) we mean that the geometrical shape of the vibration does not change during a lattice translation, that is, the displacement of a certain atom is the same in every unit cell. Therefore, only those vibrations are infrared- or Raman-active where corresponding atoms vibrate in phase in all unit cells.

For very small crystals or short polymer chains such as normal hydrocarbons the selection rules do not hold strictly since the boundary conditions are not fulfilled. All vibrations are infrared- and Raman-active, but it turns out that most of the absorption lines are weak except for the ones corresponding to totally symmetrical vibrations which would also be active in an infinite crystal or chain. This problem is discussed in detail in Chapter IV.

Mathematical Formulation of General Selection Rules

We now will give a mathematical formulation of the selection rules that will also be applicable to overtone and combination bands. Consider a transition between an initial state ψ_i and a final state ψ_f. This transition is infrared- or Raman-active if the product $\psi_i \psi_f^*$ belongs to the totally symmetrical representation of the translation group. The representations of the individual states are characterized by wave number vectors $\varkappa_i$ and $\varkappa_f$. According to Eq. (36) the character for the initial state ψ_i is given by

$$\exp\left[2\pi i(\varkappa_i \mathbf{T})\right] \tag{82a}$$

and for the conjugate complex wave function ψ_f^* of the final state by

$$\exp\left[-2\pi i(\varkappa_f \mathbf{T})\right] \tag{82b}$$

The product of the two characters only leads to the totally symmetric representation (all spurs are equal to unity) if

$$\exp\{2\pi i[(\varkappa_i - \varkappa_f)\mathbf{T}]\} = 1 \tag{83}$$

or if $(\varkappa_i - \varkappa_f)\mathbf{T}$ is an integer. From Eqs. (29) and (32) we find that this condition is fulfilled if

$$\varkappa_i - \varkappa_f = m_1 b_1 + m_2 b_2 + m_3 b_3 \tag{84}$$

where m_1, m_2, and m_3 can be any integer. b_i are the reciprocal fundamental lattice vectors. From Eq. (37) we find that a complete set of irreducible representations of the translation group can be given by a set of wave-number vectors with coefficients $0 \leqslant p_i/N_i < 1$, that is, by vectors within the first Brillouin zone. Two such vectors can satisfy Eq. (84) only if they are equal, therefore, $m_1 = m_2 = m_3 = 0$. The selection rule for the lattice translation group can then simply be stated as follows: A transition between the states ψ_i and ψ_f is infrared- or Raman-active if

$$\varkappa_i = \varkappa_f \tag{85}$$

that is, if the two states belong to the same representation. The selection rule for fundamentals is a special case where the ground state ψ_i is totally symmetrical and $\varkappa_i = \varkappa_f = 0$.

Combination of Lattice and Molecular Vibrations (Difference Bands)

Equation (85) also holds for any combination or overtone band. Of particular importance are the combinations of low energy lattice vibrations with molecular vibrations. We illustrate this type of transition for a one-dimensional diatomic lattice for which we just consider the longitudinal vibrations. Some properties of the one-dimensional translation group are given in Table II, where a chain segment consisting of N repeat units is considered. Each repeat unit contains two atoms ($m = 2$). The total number of longitudinal vibrations is, therefore, $2N$, that is, each representation (Table II) contains two longitudinal normal vibrations. In the next chapter we will discuss this case in some detail and show that these vibrations can be divided into two groups of N vibrations each. In one group the frequencies are very close to the vibrational frequency of the isolated diatomic molecule of the unit cell. The other group contains N lattice vibrations, usually of much lower frequency. These two groups of vibrations constitute the optical and the acoustical frequency branches of the system if the frequencies are plotted against the wave-number vector as shown for an example in Fig. 8a. The $2N$ energy levels of the system are shown in Fig. 8b for two arbitrarily chosen examples where $N = 6$ and $N = 9$, respectively. Each energy level belongs to a representation of the translation group, that is, to a $\varkappa$ which assumes the values

$$\varkappa = \frac{p}{Nt} \tag{86}$$

where $p = 0, 1, 2, \ldots, N-1$, and t is the lattice repeat distance. The value of p is marked for each energy level in Fig. 8b. According to Eq. (85) only transitions between levels of the same $\varkappa$ (or p value) are infrared- or Raman-active. For $\varkappa = 0$ we have a transition between the ground state and the excited state for which the corresponding atoms in each unit cell move in phase. The frequency ω_0 of this transition corresponds to a fundamental vibration of the system which usually causes a strong absorption while all the other transitions for which $\varkappa \neq 0$ are difference lines between molecular and lattice vibrations. These are expected to

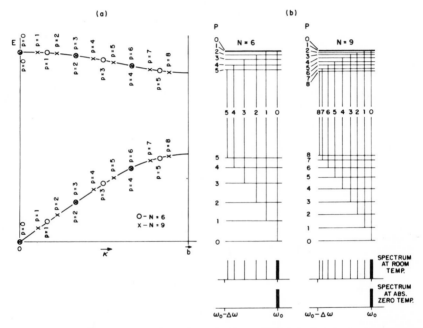

FIG. 8. (a) Frequency branches for a one-dimensional diatomic lattice (see Fig. 5, Chapter III). The energy is plotted against the wave-number vector $\varkappa$. Chain segments of six and nine repeat units, respectively, were chosen arbitrarily. (b) Energy level diagram and allowed transitions for difference bands. They are infrared- as well as Raman-active.

be quite weak in most cases (Ref. 11). We find five such lines for $N = 6$ and eight for $N = 9$. At this point we want to emphasize again that the choice of N is arbitrary. We, therefore, would not expect to observe a definite number of difference lines but rather a continuous absorption band between the frequencies $\omega_0 - \Delta\omega$ and ω_0. N can be chosen to be very large so that the individual lines can no longer be identified.

Temperature Dependence of Transitions Involving Lattice Vibrations

It is only possible to observe an absorption of a certain line if the lower state of the transition is populated. The population of the energy states is determined by a

Boltzmann distribution function which depends upon the energy of the state and the temperature of the system. Lattice vibrational states usually range in energy from 0 to about 100 (200) cm^{-1}. This means that at room temperature these states are quite highly populated, but if we measure the spectrum at liquid nitrogen or, even better, at liquid helium temperature only the very lowest lattice vibrational

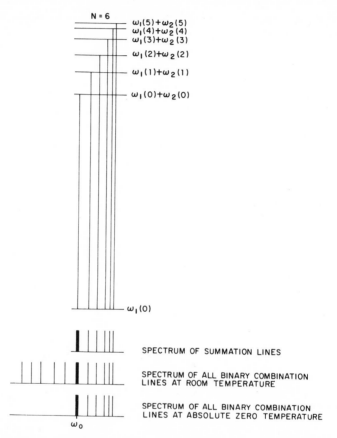

FIG. 9. Energy level diagram and allowed transitions of binary summation vibrations for a diatomic one-dimensional lattice for a chain segment consisting of six repeat units.

states are populated. In this case only the transitions for which $\varkappa$ is zero or very close to zero are observed in the spectrum. The overall effect is a marked sharpening of the absorption band by going to lower temperatures. In the particular case discussed in Fig. 8 we also would expect to find a shift of the apparent absorption maximum toward higher frequencies (Ref. 15). This characteristic temperature sensitivity of the difference bands involving low energy lattice vibration helps to identify them as such.

Summation Bands and Overtones

Binary summation bands of the frequency $\omega_1 + \omega_2$ will also cause a broadening of the absorption band of a fundamental vibration if ω_1 belongs to the acoustical and ω_2 to the optical branch. The initial state of such a transition is the ground state of the system and is, therefore, totally symmetric ($\varkappa = 0$). A transition is infrared- or Raman-active if the final state is also totally symmetric so that Eq. (85) is satisfied. The symmetry of the upper state is determined by the symmetry of the vibrations ω_1 and ω_2. They have to belong to the same representation of the translation group (same $\varkappa$ value) for their combination state to be totally symmetric. These states are shown in Fig. 9 for the diatomic lattice of Fig. 8. All the active transitions, also shown in the same figure, involve the ground state and are, therefore, not temperature sensitive.

Finite Crystals

Finally we discuss the spectrum of a finite crystal (finite chain). This problem is treated extensively in Chapter IV but we still want to mention it here because of the temperature sensitivity of the expected absorption lines. Since we now deal

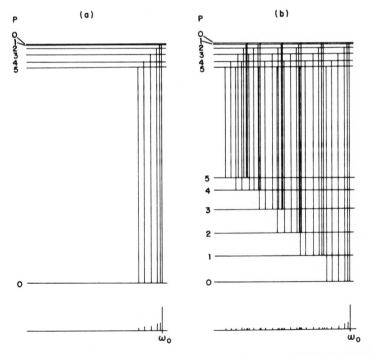

Fig. 10. Energy level diagram and allowed transitions for a one-dimensional diatomic *finite* lattice of six repeat units. (a) At absolute zero temperature—one strong line and five weak ones of decreasing intensity with decreasing frequency. (b) At room temperature same as (a) but 30 additional weak lines, the intensity of which is not known.

with an isolated finite segment the Born boundary conditions do not hold and the selection rules for infinitely long chains cannot strictly be applied. As an example, we consider again a chain as in Fig. 8 consisting of six repeat units. At absolute

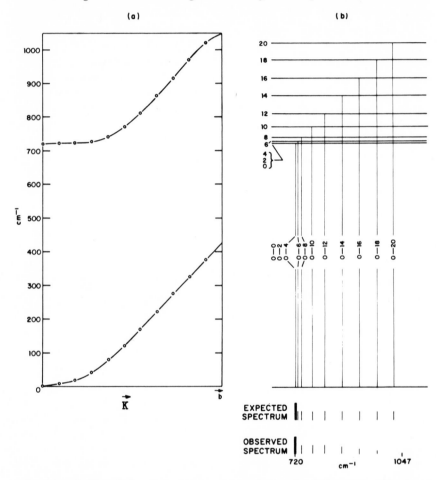

Fig. 11. $C_{24}H_{50}$ chain. (a) I, frequency branch for the CH_2-rocking vibrations. II, acoustical frequency branch for the perpendicular skeletal vibrations. (b) Energy level diagram and absorption spectrum at absolute zero temperature.

zero temperature only the ground state is populated. We will show in Chapter IV that not only the 0-0 transition but also the other ones, namely, 0-1, 0-2, etc., are active with decreasing intensity by going from the 0-0 to the 0-5 transition. Such a low temperature spectrum is shown schematically in Fig. 10a.† At room tem-

† If the lattice has a higher symmetry (space group) other selection rules may hold which, for example, might cause alternating lines to have zero intensity.

perature the low energy lattice states are populated as well as the ground state so that all possible transitions become infrared active as shown in Fig. 10b but the 0-0 transition will still be by far the strongest one. The weak lines (in this case

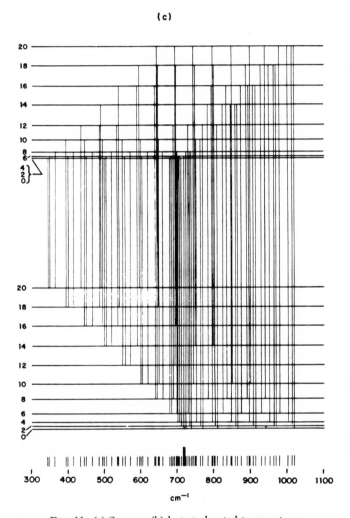

FIG. 11. (c) Same as (b) but at elevated temperature.

there are 35) can usually not be observed as individual absorption lines, but rather as a broad smeared out absorption band. In many cases such a multiplicity of lines just appears as a continuous absorption which has the effect of somewhat increasing the background in an absorption spectrum. At low temperature we would expect to be able to observe the five individual lines of Fig. 10a. An example of such a spectrum is shown in Fig. 10a of Chapter IV for the CH_2-rocking vibrations of $C_{24}H_{50}$ at $-160°C$ (Ref. 30). This chain consists of 22 CH_2 groups. It will be treated in Chapter IV as a set of 22 coupled oscillators. The vibrational energy is given by the frequency branches in Fig. 11a as a function of the wave-number vector $\varkappa$. The optical branch is obtained from Fig. 10a in Chapter IV and the acoustical branch is the one for perpendicular skeletal vibrations of a polyethylene chain given in Fig. 13a of Chapter III. At very low temperature only transitions from the ground state will be allowed. It turns out (see Chapter IV) that for symmetry reasons only 11 (instead of 22) lines are infrared-active. (Another way to explain this is by the fact that the translational repeat unit in a hydrocarbon chain contains two CH_2 groups and the $C_{24}H_{50}$ molecule, therefore is a system for which $N = 11$.) The energy levels and the low temperature transitions are shown in Fig. 11b. The expected absorption spectrum (11 lines) is also shown with the 0-0 transition being the strongest one. Experimentally seven lines are observed. At higher temperature the total number of possible transitions is 121 as shown in Fig. 11c. Only the strongest line is observed in an actual spectrum while the weak ones are too close together to show up as individual lines. In this particular example we only treated an individual chain molecule, but if we consider neighboring chains in a crystal, there are additional lattice vibrations leading to low energy levels and the expected spectrum is much more complex.

F. SPACE GROUPS

General Discussion of Selection Rules

From the previous sections we recall that a space group is composed of the lattice translation group and additional symmetry elements, such as rotations, reflections, etc. The translation group is, therefore, a subgroup of the space group. The selection rules for these two symmetry groups are very closely related. To understand this relation consider any vibrating system with certain symmetry elements forming a group, which determines the vibrational selection rules. Now we assume that the symmetry of the system is lowered so that it can be described by a subgroup of the original group. In this case the selection rules are less strict, and in general more vibrations are infrared- or Raman-active, but it is important to note that the selection rules for the subgroup also hold for the original group of higher symmetry. Therefore, the selection rules for the translation group, discussed in the foregoing subsection, also apply to any space group. They are necessary, but not sufficient, that is fewer vibrations are infrared- or Raman-active for the space group than for the translation group.

First we consider fundamental vibrations. From the translation group analysis we know that only the vibrations that belong to the totally symmetric representation $\Gamma^{(0)}$ are infrared- and Raman-active. These are only $3m$ vibrations, where m is the number of atoms per unit cell. If the lattice has space group symmetry the same $3m$ vibrations will be divided among the so-called factor group representations of the space group. These are the representations that reduce to $\Gamma^{(0)}$ in the translation group if all the nontranslational symmetry elements of the space group are dropped. This problem was discussed in Section 3B. For a space group the $3m$ factor group vibrations are potentially infrared- or Raman-active. The selection rules are determined exactly as in the case of point groups, where only those of the $3m$ vibrations are infrared-active which belong to the same representation as one of the three components of a vector. Analogous rules hold for the Raman activity where we have to consider the six components of the polarizability tensor as discussed in detail in Sections 4A to 4C. The spectral activities for the fundamentals are given for two examples in Tables IV and V.

The selection rules for overtone and combination lines can again be discussed in the light of the selection rules for the translation group. To do this we need some information about the irreducible representations of space groups. This problem is treated in the next paragraph.

Irreducible Representations of Space Groups

Space groups have been described in previous sections where it was shown that the order of a three-dimensional finite space group (fulfilling the Born boundary conditions) is $N_1 N_2 N_3 H$, where H is the order of the factor group. A space group symmetry operation is in general a combination of a translation- and a point-symmetry element. A representation of the space group consists, therefore, of matrices which are products of matrices of translation group and point group representations (the situation is slightly more complicated if the space group contains screw rotations and glide reflections) (Refs. 25, 26). Some space group representations are only one-dimensional, others can be as high as H-dimensional. Winston and Halford (Ref. 37) have shown that the spur $\chi_j^{(\varkappa)}$ of a matrix for a space group representation is given by the following expression (it is the sum over the diagonal elements of the representation matrix):

$$\chi_j^{(\varkappa)}(R, \mathbf{T} + \boldsymbol{\tau}_R) = \sum_{\substack{\varkappa \text{ in } \{\varkappa\} \\ (H/k \text{ elements})}} \exp\left[2\pi i \varkappa(\mathbf{T} + \boldsymbol{\tau}_R)\right] \chi_j^{(K)}(R)\, \delta_{R\varkappa, \varkappa} \quad (87)$$

where $(R, \mathbf{T} + \boldsymbol{\tau}_R)$ is a space group element. $\mathbf{T}$ is a lattice translation vector (element of translation group), R is a point group element and $\boldsymbol{\tau}_R$ is a fractional lattice translation which only differs from zero for space groups containing either screw rotations or glide reflections. For the screw rotation in polyethylene (Table V, Fig. 4), for example, R is a rotation of $180°$ and $\boldsymbol{\tau}_R$ is a translation by one half of the lattice unit-vector. The space group representations are H/k-dimensional matrices. Their diagonal elements are products of exponential functions (these are essentially the representations of the translation group) and the spurs $\chi_j^{(K)}(R)$ for the jth irreducible representation of the point group K, the significance of which we now will explain. Consider a wave number vector $\varkappa$ and apply to it all the H point-symmetry operations R of the space group. In the most general case a set $\{\varkappa\}$ of H vectors is created by this procedure, but in some special cases, where the vector $\boldsymbol{\varkappa}$ coincides with point-symmetry elements, it can be invariant under the corresponding symmetry operations R. These symmetry operations form a group K of the order k and the set of wave-number vectors $\{\boldsymbol{\varkappa}\}$ only contains H/k different vectors. The factor $\delta_{R\varkappa,\varkappa}$ in Eq. (87) means that the diagonal elements of a space group matrix are zero if $\boldsymbol{\varkappa}$ is not invariant under the operation R; that is, if R is not a member of K.

To summarize, we can say that an irreducible space group representation can be characterized by a wave-number vector $\boldsymbol{\varkappa}$ and an irreducible representation of a point group K. For the trivial case of the translation group the only point symmetry element R is unity E, and the point group K is the trivial group C_1. Equation (87) then reduces to Eq. (36). Another trivial set of representations are the factor group representations. They are obtained from Eq. (87) if the wave-number vector is equal to zero. Every point symmetry operation R leaves this trivial vector invariant. The group K is, therefore, identical with the factor group $(k = H)$ and the set $\{\boldsymbol{\varkappa}\}$ consists of the zero vector only. Equation (87) then reduces to

$$\chi_j^{(0)}(R, \mathbf{T} + \boldsymbol{\tau}_R) = \chi_j^{(\text{f.g.})}(R) \tag{88}$$

It is important to note that the value of $\chi_j^{(0)}$ is independent of the translational part of the space group element. In other words, $\chi_j^{(0)}$ is the same for every element in a coset. This is shown explicitly in Table X for the one-dimensional line group C_{2v}; $\boldsymbol{\tau}_R$ is equal to zero for this example. The factor group representations are usually presented in an abbreviated form as shown earlier in Tables IV and V.

The Most General Form of Space Group Selection Rules

We now are in a position to apply the selection rules derived in the previous subsections to systems with space group symmetry. We recall that a transition between an initial state i and a final state f is allowed only if the product of the two wave functions $\psi_i\,\psi_f^*$ has the symmetry of an activity representation; that is, a representation which contains a component of the dipole moment vector or the polarizability tensor. In the case of the translation group all these components belong to the representation for which $\boldsymbol{\varkappa} = 0$. The same is true for space groups, but now there are several representations with $\boldsymbol{\varkappa} = 0$, namely, the factor group representations. In general only some of them are activity representations. They can be obtained from an analysis of point groups (see e.g., Ref. 11) since the factor group of a space group is always isomorphous with a point group. Consider now the two characters of the form of Eq. (87) for the initial and final states. Two conditions have to be fulfilled for their product to be an activity representation. The first one is Eq. (85) which is the selection rule for the pure translation group. It says that the product has to contain a factor group representation. For space groups this condition is necessary but not sufficient. The second and more restrictive condition says that at least one of the factor group representations contained in the product has to be an activity representation. Whether or not this

TABLE X

Explicit Factor Group Representations for the Line Group C_{2v}

	Space group elements $(R, \mathbf{T})$ arranged in four cosets											
	(E,T_1)	(E,T_2)	$\cdots$	(E,T_N)	(C_2',T_1)	(C_2',T_1)	$\cdots$	$(C_2'T_N)$	(σ_h,T_1)	(σ_h,T_2)	$\cdots$	(σ_h,T_N)	(σ_v'',T_1)	$(\sigma_v''T_2)$	$\cdots$	(σ_v'',T_N)
$\chi_1^{(0)}(R,\mathbf{T})$	1	1	$\vdots$	1	1	1	$\vdots$	1	1	1	$\vdots$	1	1	1	$\vdots$	1
$\chi_2^{(0)}(R,\mathbf{T})$	1	1	$\vdots$	1	1	1	$\vdots$	1	-1	-1	$\vdots$	-1	-1	-1	$\vdots$	-1
$\chi_3^{(0)}(R,\mathbf{T})$	1	1	$\vdots$	1	-1	-1	$\vdots$	-1	-1	-1	$\vdots$	-1	1	1	$\vdots$	1
$\chi_4^{(0)}(R,\mathbf{T})$	1	1	$\vdots$	1	-1	-1	$\vdots$	-1	1	1	$\vdots$	1	-1	-1	$\vdots$	-1

is the case can be decided with the help of an equation also derived by Winston and Halford (Ref. 37). It gives the number of times, $a_j^{\text{f.g.}}$, the jth factor group representation appears in the product representation under the condition that Eq. (85) is fulfilled:

$$a_j^{\text{f.g.}} = \frac{1}{k} \sum_{R \text{ in } (K)} \chi_i^{(K)}(R) \, \chi_f^{*(K)}(R) \, \chi_j^{\text{f.g.}}(R) \tag{89}$$

$\chi_i^{(K)}$ and $\chi_f^{(K)}$ are the characters of the subgroups K associated with the initial and final state ψ_i and ψ_f respectively, and k is the order of the group K as explained after Eq. (87). $\chi_j^{\text{f.g.}}$ is the character of the jth factor group representation, listed in the character Tables IV and V for two examples. A transition is infrared- or Raman-active if $a_j^{\text{f.g.}}$ is larger than zero and if the jth representation is an activity representation.

Let us now consider Eq. (89) for three trivial but important cases. For the pure translation group the factor group as well as the group K reduce to the group C_1 consisting of the unit element only ($k = 1$), and $a_j^{\text{f.g.}}$ becomes equal to one ($j = 1$). This means that every transition for which Eq. (85) holds is infrared- or Raman-active for a pure translation group, a result derived previously. In the second trivial case we apply Eq. (89) to fundamental vibrations. Since the wave number vector ($\varkappa = 0$) is invariant for any point group operation R the group K becomes isomorphous with the factor group and k is equal to H. Furthermore, the initial state (ground state) is totally symmetric so that $\chi_i(R) = 1$. Equation (89) then reduces to

$$a_j^{\text{f.g.}} = \frac{1}{H} \sum_{R \text{ in f.g.}} \chi_f^{*\,\text{f.g.}}(R) \, \chi_j^{\text{f.g.}}(R) \tag{90}$$

and with Eq. (26) we obtain

$$a_j^{\text{f.g.}} = \delta_{fj} \tag{91}$$

This confirms the previously derived result that a fundamental vibration can only be infrared- or Raman-active ($a_j^{\text{f.g.}} \neq 0$) if the final state f (excited state) belongs to an activity representation j. The third trivial case we consider is the one where the vector $\varkappa$ does not lie on any symmetry element and is, therefore, not invariant to any symmetry operation. In such a case the group K reduces to the trivial point group C_1, consisting of the unit element E, and Eq. (89) becomes

$$a_j^{\text{f.g.}} = \chi_j^{\text{f.g.}}(E) \tag{92}$$

The right-hand side of Eq. (92) is the spur of the unit matrix. It is always one or larger than one regardless of the value of j, which means that the product representation of the initial and final states contains all factor group representations, including all activity representations. Every such transition will, therefore, be infrared- as well as Raman-active. This finding is important for the selection rules of overtones and combination bands discussed in the following paragraphs.

Selection Rules for Overtones and Combination Bands

The selection rules for any transition are fully determined by Eqs. (85) and (89) if the representations of all the energy states involved are known. Here we mention a few examples of overtone and combination bands. For a first overtone or binary summation band the initial state ψ_i is the ground state (totally symmetric). The final state ψ_f is a combination of the first excited state of two fundamentals ν_1 and ν_2. The representation of ψ_f is the product of the representations for the two states ψ_{ν_1} and ψ_{ν_2}, determined by Eq. (89) where ψ_i and ψ_f are replaced by ψ_{ν_1} and ψ_{ν_2}, respectively. If $\psi_{\nu_1}\psi_{\nu_2}^*$ belongs to an activity representation (i.e. if the respective $a_i^{f.g.}$ value is $\neq 0$) the combination band is allowed. For first overtones where $\nu_1 = \nu_2$ the representation of the product $\psi_{\nu_1}\psi_{\nu_1}^*$ always contains the totally symmetric representation. This means that the first overtone of any vibration is always Raman-active while it may or may not be infrared-active. Infrared spectra of overtone and summation vibrations have been observed for a number of polymers while the corresponding Raman spectra are much more difficult to obtain, although Nielsen and Woollett (Ref. 23) observed some lines in the Raman effect for polyethylene which might be assigned to overtone and combination vibrations.

Difference bands are of particular importance, especially the ones that involve low energy lattice vibrations which are thermally excited at room temperature. These so-called hot bands have been described earlier for lattices with pure translational symmetry where the temperature dependence and the selection rules were discussed in detail. Restrictions in addition to the selection rule of Eq. (85) are again determined by Eq. (89).

Finally we point out that for a three-dimensional lattice the number of active overtone and combination vibrations is very large (Ref. 37) since a transition between any two states that have the same wave number vector $\varkappa$ is active if $\varkappa$ does not lie on a symmetry element. This was shown in the discussion of Eq. (92). The number of such vectors is almost equal to the number of unit cells $N_1 N_2 N_3$ in the crystal considered. The important point to note is that most combination and overtones are, therefore, infrared- and Raman-active for a three-dimensional lattice. It is, for example, possible that for an inactive factor group vibration the combination bands with lattice vibrations are active. This can lead to broad absorptions near the frequency of the inactive fundamental vibration.

G. Site Groups

Halford (Ref. 10) has introduced the concept of the "site-group" which allows a simplification of the treatment in many practical cases. It is based on the fact that quite often the spectra of individual molecules (e.g. in a dilute solution) and a crystal formed by the same molecules are similar. Differences such as splittings and shifts of bands can then be considered as perturbations because the interaction between molecules in a crystal is weak compared with the forces within the molecule itself. The selection rules, on the other hand, do not depend upon the force constants but solely upon the symmetry. Therefore, Halford suggested to analyze the vibrations of an individual molecule but to determine their activity by the symmetry of the surrounding force field of the crystal. This point-symmetry is characterized by the so-called "site-group" which is a subgroup of the space group. In the case of chain molecules the interaction between repeat units is very strong, since they are usually connected by covalent bonds. Therefore, the site-group approximation becomes unrealistic. To overcome this difficulty Tobin (Ref. 35) has used a different approach by introducing the concept of the line group, where in a first approximation the chain is treated as a one-dimensional crystal. Such a treatment can give the main features of the spectrum while for the finer points (e.g. band splittings) a space group analysis is required. Furthermore, the selection rules might be somewhat different in a line group than in a space group treatment as we will discuss in the case of polyethylene in Section 6.

5. Summary

Consider a three-dimensional crystal with $N_1 N_2 N_3$ unit cells and m atoms per unit cell. Such a crystal has $3mN_1 N_2 N_3$ normal vibrations. They can be equally divided among the $N_1 N_2 N_3$ representations of the translation group, each of which is characterized by a wave-number vector $\varkappa$ [Eq. (36)]. Only the $3m$ vibrations for which $\varkappa = 0$ are infrared- and Raman-active as fundamentals. These are the vibrations for which corresponding atoms in each unit cell move in phase. For a crystal with space group symmetry the selection rules are even stricter. In this case there are several representations for which $\varkappa = 0$, the so-called factor group representations. They contain a total of $3m$ vibrations but only those are active that belong to an activity representation.

For overtones and combination bands the selection rules have to be given in their most general form. For translation group symmetry a transition is allowed if the initial and final states correspond to the same wave-number vector [see Eq. (85)]. In general the number of allowed transitions is very large. For space group symmetry the selection rules are somewhat stricter, but it turns out that most of the transitions allowed under translational symmetry are still active. Transitions between low frequency lattice vibrations and "molecular vibrations" are of particular importance since they may determine the shape and width of absorption bands. They are temperature sensitive and can, therefore, be identified with the help of low temperature spectra.

6. Illustrations

A. POLYETHYLENE

Polyethylene in its purest form has a very simple structure and therefore an absorption spectrum relatively easy to analyze. In this section we will summarize the major features of such an analysis for the fundamental vibrations.

Structure and Symmetry

In the crystalline state the carbon backbone of a polyethylene molecule is a planar zig-zag chain as shown in Fig. 4. The repeat unit consists of two CH_2 groups. The symmetry properties of this molecule are discussed in Section 2 in

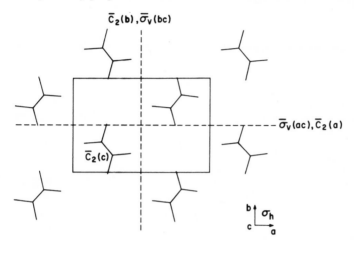

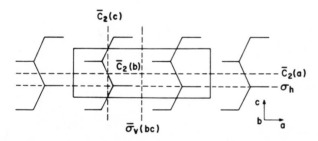

Fig. 12. Orthorhombic unit cell (shown as solid line) of polyethylene after Bunn (Ref. 6). The molecular chain axis is the c-axis of the crystal. Space group $P \dfrac{2_1}{n} \dfrac{2_1}{a} \dfrac{2_1}{m}$; $a = 7.40\,\text{Å}$, $b = 4.93\,\text{Å}$, $c = 2.534\,\text{Å}$.

terms of the one-dimensional space group (line group) V_h. The character table for this group as well as a classification of the factor group normal vibrations are given in Table V and the geometrical form of the vibrations is shown in Fig. 7 (Refs. 17, 18, 23, 35).

In practice, we do not deal with isolated chains but rather with a highly crystalline polymer [crystallinity may be as high as 80 to 90% (Ref. 21)]. We therefore will first discuss the crystal structure of polyethylene and show that the main features of the spectrum can be interpreted in terms of a pure crystal spectrum.

The crystal structure of polyethylene has been determined by Bunn (Ref. 6). He showed by X-ray diffraction that the structure is the same as in some n-paraffins (Ref. 20). The unit cell is orthorhombic with the dimensions shown in Fig. 12. It contains four methylene groups. From Fig. 12 it is obvious that two molecular chains pass through this unit cell.

The symmetry of the crystal is characterized by a space group. Its designation is $P\dfrac{2_1}{n}\dfrac{2_1}{a}\dfrac{2_1}{m}$, using the nomenclature of Buerger (Ref. 5). The factor group of this space group is of the order eight and its elements are shown in Fig. 12. They can be described as follows: Unity E, three mutually perpendicular twofold screw axis $\bar{C}_2(a)$, $\bar{C}_2(b)$, and $\bar{C}_2(c)$, a glide plane $\bar{\sigma}_v(bc)$ with a translational component of $b/2 + c/2$, a glide plane $\bar{\sigma}_v(ac)$ with a translational component $a/2$, a plane $\sigma_h(ab)$ in the plane of the CH_2 group, and a center of inversion i on the C—C bond center. This factor group is isomorphous with the point group V_h and therefore has the same character table as shown in Table XI. The fact that the factor group of the space group and the line group are isomorphous with the *same* point group is accidental and is not generally true for other polymers. We now will turn our attention to the normal vibrations of a polyethylene crystal.

Normal Vibrations and Their Spectral Activity

The unit cell contains four carbon and eight hydrogen atoms. The number of factor group fundamental vibrations, $3N$, is therefore 12 for the carbon skeleton and 36 for the whole molecule. These 36 vibrations are also called the unit cell vibrations since all the unit cells in the crystal vibrate in the same phase. We now will determine how these vibrations can be assigned to the eight species shown in Table XI. The number of vibrations corresponding to the lth factor group representation (species) is given by Eq. (73). For polyethylene, $H = 8$, the character $\chi^{(l)}(F_i)$ is listed in Table XI for the factor group symmetry elements F_i. The values of $u(\text{coset})$ and $2\cos\varphi \pm 1$ are given in Table XII. With the numerical values of Tables XI and XII, Eq. (73) becomes

$$a_1 = \tfrac{1}{8}(+1 \times 4 \times 3 + 1 \times 4 \times 1) = 2$$
$$a_2 = \tfrac{1}{8}(+1 \times 4 \times 3 - 1 \times 4 \times 1) = 1$$

etc.

for the carbon skeleton vibrations, and

$$a_1 = \tfrac{1}{8}(+1 \times 12 \times 3 + 1 \times 12 \times 1) = 6$$
$$a_2 = \tfrac{1}{8}(+1 \times 12 \times 3 - 1 \times 12 \times 1) = 3$$

etc.

for the vibrations including the hydrogen atoms. The results of these computations are listed in columns 10 and 11 of Table XI. Three of the 36 unit cell vibrations are pure translations of the whole lattice. They belong to the species B_{1u}^S, B_{2u}^S,

TABLE XI

Space Group $P\,\dfrac{2_1}{n}\,\dfrac{2_1}{a}\,\dfrac{2_1}{m}$ for Crystalline Polyethylene;

The Factor Group Is Isomorphous with the Point Group V_h

| Species | Factor group symmetry elements F_i | | | | | | | | a_i | a_i skeleton | Pure translations | Spectral activity | |
	E	$\sigma_h(ab)$	$\bar{\sigma}_v(bc)$	$\bar{\sigma}_v(ac)$	$\bar{C}_2(c)$	$\bar{C}_2(a)$	$\bar{C}_2(b)$	i				IR	Raman
A_g^S	+1	+1	+1	+1	+1	+1	+1	+1	6	2		i	$\alpha_{aa},\ \alpha_{bb},\ \alpha_{cc}$
A_u^S	+1	−1	−1	−1	+1	+1	+1	−1	3	1		i	i
B_{1g}^S	+1	+1	−1	−1	+1	−1	−1	+1	6	2		i	α_{ab}
B_{1u}^S	+1	−1	+1	+1	+1	−1	−1	−1	3	1	T_c	M_c	i
B_{2g}^S	+1	+1	−1	+1	−1	−1	+1	+1	3	1		i	α_{ac}
B_{2u}^S	+1	−1	+1	−1	−1	−1	+1	−1	6	2	T_b	M_b	i
B_{3g}^S	+1	+1	+1	−1	−1	+1	−1	+1	3	1		i	α_{bc}
B_{3u}^S	+1	+1	−1	+1	−1	+1	−1	−1	6	2	T_a	M_a	i

and B_{3u}^S, respectively. No free rotations of the crystal as a whole are possible because the space group theory either assumes an infinite lattice or a lattice with the Born boundary condition (see Section 2).

The spectral activity of the unit cell vibrations is also given in Table XI. It is determined solely by the point group isomorphous with the factor group and is therefore analogous to the activity shown in Table V. Eighteen vibrations are Raman-active, 15 vibrations (including three translations) are infrared-active, and

TABLE XII

ANGLE OF ROTATION φ, $2 \cos \varphi \pm 1$, AND u(coset) (= NUMBER OF ATOMS PER UNIT CELL LEFT INVARIANT BY PARTICULAR SYMMETRY OPERATION) FOR CRYSTALLINE POLYETHYLENE

Factor group symmetry elements	φ (deg)	$2 \cos \varphi \pm 1^a$	u(coset) Carbon skeleton	u(coset) Whole molecule
E	0	3	4	12
$\sigma_h(ab)$	0	1	4	12
$\bar{\sigma}_v(bc)$	0	1	0	0
$\bar{\sigma}_v(ac)$	0	1	0	0
$\bar{C}_2(c)$	180	−1	0	0
$\bar{C}_2(a)$	180	−1	0	0
$\bar{C}_2(b)$	180	−1	0	0
i	180	−3	0	0

^{a} Plus (+) sign for rotation, minus (−) sign for rotation reflection.

the 3 vibrations of species A_u are neither infrared- nor Raman-active. We further note that a given vibration cannot be both infrared- *and* Raman-active. This is the well-known mutual exclusion rule that applies whenever a system has a center of symmetry.

Comparison of Line Group and Space Group Analysis

In previous sections, we have shown that in a first approximation to a vibrational analysis individual polymer chains can be treated as one-dimensional crystals characterized by a line group. In a more rigorous treatment intermolecular forces are not neglected, which leads to the space group analysis. The intermolecular interaction does not drastically influence the spectrum but may cause a splitting of absorption bands as we shall show in some detail in the next few paragraphs.

According to the line group treatment, the individual polyethylene chain has 18 factor group normal vibrations (14 of them are genuine, see Table V). The molecular motions for these vibrations are shown in Fig. 7. A space group analysis leads to twice as many factor group normal vibrations, so that one vibration of the line group corresponds to two vibrations of the space group. These two vibrations would have the same frequency if intremolecular interaction could be neglected, but in reality a doublet splitting is observed at least for some of the absorption bands. As an illustration we consider the now classic example of the CH_2-rocking vibration (Ref. 33). For an individual chain, the geometrical motion is shown in Fig. 13a. It belongs to the line group species B_{2u}^L because it is symmetric with

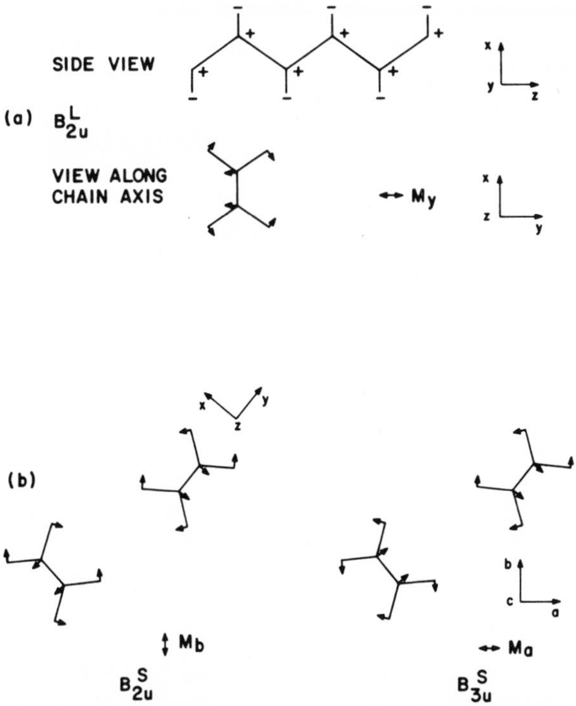

FIG. 13. CH_2 rocking vibration in polyethylene. (a) Individual chain characterized by a line group, vibration belongs to species B_{2u}^L with a transition moment in the y-direction. (b) Crystal characterized by space group. Two vibrations belong to species B_{2u}^S and B_{3u}^S with a transition moment in the b- and a- directions, respectively.

respect to the line group elements E, σ_h, $\bar{\sigma}_v'$, C_2'' and antisymmetric with respect to σ_v'', $\bar{C}_2$, C_2', i (see character table Table V). Its transition moment is in the y-direction. The corresponding two vibrations in a crystal are shown in Fig. 13b where the rocking motions for both chains of the unit cell are in-phase and out-of-phase, respectively (the definition of which motion corresponds to the in-phase or out-of-phase vibration is arbitrary). The two vibrations occur at 720 and 731 cm^{-1}. They belong to two different species of the space group. One belongs to B_{2u}^S and absorbs in the b-direction of the crystal, the other one belongs to B_{3u}^S and has its transition moment in the a-direction as shown by Krimm (Ref. 16), by observing the dichroism in the spectrum of a single crystal of n-$C_{36}H_{74}$ with the radiation beam parallel to the c-axis of the crystal. The amount of splitting of this band depends on the degree of intermolecular interaction which may vary somewhat with temperature as shown by Novak (Ref. 24).

In general, a vibration of a line group species corresponds to two vibrations of two different space group species. The correspondence of species can be found by the following consideration. Four of the symmetry elements are the same for the line group and space group. They are E, σ_h, $\bar{C}_2$, and i (they form the site-group line group, Ref. 35). The behavior of a given vibration, that is, whether it is symmetric or antisymmetric with respect to one of these symmetry elements, should therefore be the same for a line group species and a space group species.

TABLE XIII

CORRESPONDING SPECIES FOR THE FACTOR GROUP OF THE
LINE GROUP AND THE SPACE GROUP OF POLYETHYLENE

Line group	Space group
A_g^L	A_g^S, B_{1g}^S
A_u^L	A_u^S, B_{1u}^S
B_{1g}^L	B_{1g}^S, A_g^S
B_{1u}^L	B_{1u}^S, A_u^S
B_{2g}^L	B_{2g}^S, B_{3g}^S
B_{2u}^L	B_{2u}^S, B_{3u}^S
B_{3g}^L	B_{3g}^S, B_{2g}^S
B_{3u}^L	B_{3u}^S, B_{2u}^S

For example, for a vibration of line group species B_{2u}^L we find the entries in the character table (Table V) to be $+1$, $+1$, -1, and -1 for the four symmetry elements that are common to space and line groups. In the character table for the space group (Table XI) we find two species, namely, B_{2u}^S and B_{3u}^S that have the same character for these four symmetry elements. Table XIII gives the correspondence of all the species of the two groups (see also Ref. 35, Fig. 2).

From the foregoing discussion it is clear that the 36 unit cell vibrations are the in-phase and out-of-phase modes of the 18 vibrations of a single chain (Fig. 7). This is shown in Fig. 14 for the carbon skeleton vibrations. A description of all unit cell vibrational modes has been given by Krimm, Liang, and Sutherland (Ref. 18). The 12 unit cell vibrations in Fig. 14 can be classified as four carbon skeletal vibrations, two rotational lattice vibrations, three translational lattice vibrations, and three pure translations.

Infrared spectra of linear polyethylene are shown in Figs. 15a and b. The assignment of the observed infrared- and Raman-active fundamental vibrations is given in Table XIV according to Refs. 18, 22, 23, 38. There seems to be no doubt in the assignment of the two infrared- and two Raman-active CH_2-stretching vibrations in the 2900 cm^{-1} region, the CH_2-bending doublet at about 1470 cm^{-1}, and CH_2-rocking doublet in the 720 cm^{-1} region. The assignment of the two

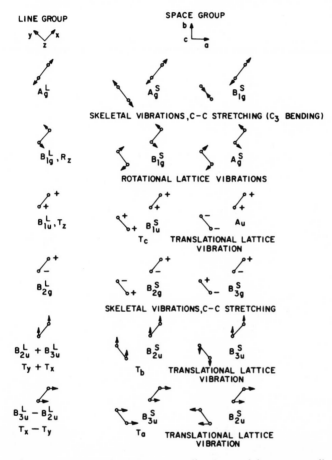

FIG. 14. Geometrical form of the six line group vibrations and the corresponding 12 space group vibrations involving motions of the carbon skeleton of polyethylene. Shown are projections in the direction of the chain axis.

CH_2-wagging vibrations at 1176 and 1415 cm^{-1} seems to be quite reasonable because they occur at the edges of the CH_2-wagging progression band series (1175–1415 cm^{-1}) observed for normal hydrocarbons (Ref. 31, see also Chapter IV, Section 4B). For an infinite chain we would expect only the lowest frequency band of the series to be infrared-active (Chapter IV) while the highest frequency band is expected to be Raman-active. This is in good agreement with the assignment in

TABLE XIV

LINE GROUP AND SPACE GROUP FUNDAMENTAL VIBRATIONS OF POLYETHYLENE

	Line group			Space group					Remarks
		Activity[a]			Activity		Observed spectra, frequency (cm^{-1})		
Species	Characteristic vibration	IR	Raman	Species	IR	Raman	IR	Raman	
A_g	C—C stretching (C_3 bending) perpendicular to chain	i	p	A_g	i	p		} 1061	~1070 computed value Fig. 12, Chapter III.
				B_{1g}	i	d			
	CH_2 symmetric stretching	i	p	A_g	i	p		} 2848	
				B_{1g}	i	d			
	CH_2 bending	i	p	A_g	i	p			Raman line expected around 1640 cm^{-1} (See Chapter IV, 4C).
				B_{1g}	i	d			
A_u	CH_2 twisting	i		A_u	i	i			For finite chain with polar end group progression band observed 1180–1322 (See Chapter IV).
				B_{1u}	M_c	i	1306[b]		
B_{1g}	Rotation around chain axis	i		A_g	i	d			Rotational lattice vibration, very low frequency expected.
				B_{1g}	i	d			
	CH_2 rocking	i	d	A_g	i	p		} 1168	
				B_{1g}	i	d			
	CH_2 asymmetric stretching	i	d	A_g	i	p		} 2883	
				B_{1g}	i	d			

Species	Vibration	M	act.	B_{1u}	T_c	act.	Frequency	Remarks
	CH₂ wagging	M_z	i	A_u	i	i	1176	for translational lattice vibration. Zero frequency for pure translation. Ref. 29.
				B_{1u}	M_c	i	(1367[b]) (1352[b])	~1150 computed value, Fig. 12, Chapter III.
B_{2g}	C—C stretching		d	B_{2g}	i	d	⎫ 1131	
				B_{3g}	i	d	⎭	
	CH₂ wagging		d	B_{2g}	i	d	⎫ 1415	
				B_{3g}	i	d	⎭	
B_{2u}	Translation T_y			B_{2u}	T_b	i		Zero frequency for pure translation.
				B_{3u}	M_a	i		Very low frequency for translational lattice vibration.
	CH₂ rocking	M_y	i	B_{2u}	M_b	i	720	
				B_{3u}	M_a	i	731	
	CH₂ asymmetric stretching	M_y	i	B_{2u}	M_b	i	⎫ 2919	Split in n-paraffin with components at 2824 and 2899 (?) (Ref. 18).
				B_{3u}	M_a	i	⎭	
B_{3g}	CH₂ twisting		d	B_{2g}	i	d	⎫ 1295	See remark for A_u twisting vibration.
				B_{3g}	i	d	⎭	
B_{3u}	Translation T_z			B_{2u}	M_b	i		Very low frequency for translational lattice vibration.
				B_{3u}	T_a			Zero frequency for pure translation.
	CH₂ symmetric stretching	M_x	i	B_{2u}	M_b	i	⎫ 2851	Split in n-paraffin with components at 2850 and 2857 (Ref. 18).
				B_{3u}	M_a	i	⎭	
	CH₂ bending	M_x	i	B_{2u}	M_b	i	1463	
				B_{3u}	M_a	i	1473	

[a] Instead of listing the components of the polarizability tensor, it is shown whether the light in the Raman line is polarized (p) or depolarized (d) (see Ref. 11).

[b] Observed for amorphous polymer.

Table XIV. The assignment of the Raman line at 1295 cm^{-1} to the CH$_2$-twisting vibration is probably also correct because it is the only Raman line in the region of the CH$_2$-twisting vibration band series for fatty acids (Chapter IV, Section 4C). The progression band series of the CH$_2$-rocking vibrations for finite hydrocarbon

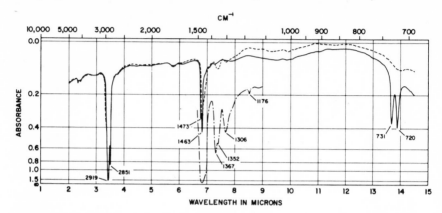

FIG. 15a. Spectra of linear polyethylene films. KEY: ———: ∼0.008 mm thick, 25°C; – – –: same film, 160°C; –·–·– 0.206 mm thick, 25°C.

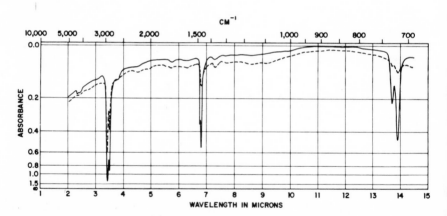

FIG. 15b. Polarization spectra of highly stretched linear polyethylene film. ∼ 0.008 mm thick. KEY: ———: electric vector perpendicular to the direction of stretch; – – –: electric vector parallel to the direction of stretch.

chains extends from 720 to about 1050 cm^{-1} and it is at this higher frequency where the Raman-active CH$_2$-rocking vibration for an infinite chain is expected. One might therefore be justified to assign the Raman line observed at 1061 cm^{-1} to this vibration and the bands at 1131 and 1168 cm^{-1} to the C—C skeletal vibrations. The latter two frequencies are well within the uncertainties of the skeletal vibration calculations of Chapter III.

B. Polyhydroxymethylene

This simple polymer with the chemical repeat unit CHOH was only recently prepared as an oriented film by Field and Schaefgen (Ref. 9). Little is known about its crystal structure and so far only a very crude interpretation of its infrared spectrum can be given. Nevertheless this spectrum is discussed in this section as an example for a rudimentary analysis.

Structure

The exact geometrical structure of the polymer in the crystalline state is not known but there is evidence from X-ray diffraction patterns of drawn samples that the repeat unit of the chain is about 2.5 Å. This indicates that the carbon

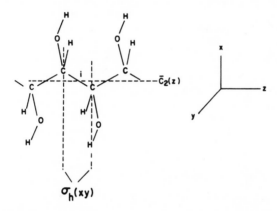

Fig. 16. Structure and symmetry of polyhydroxymethylene assuming a planar zig-zag for the chain of carbon atoms.

skeleton is a planar zig-zag chain as in polyethylene. Furthermore, the mechanism of polymerization (Ref. 9) suggests that the OH groups of adjacent carbon atoms lie on opposite sides of the zig-zag plane. The structure of such a chain is shown in Fig. 16. The z-axis of a coordinate system is parallel to the chain axis and the carbon atoms lie in the xz-plane. The CH bond and CO bond are in the xy-plane (similar to the CH_2 group of polyethylene). We further assume that the OH bond also lies in this plane with the distance between the hydroxyl and aliphatic hydrogen being the same for all CHOH groups. This assumption may not be correct and the symmetry of the chain may be lower, which would change the selection rules discussed below.

Symmetry Properties

The symmetry of the molecular structure shown in Fig. 16 can be described by a line group characterized by the following symmetry operations: a reflection at a plane of symmetry $\sigma_h(xy)$ (through the CHOH group), a screw rotation around the twofold screw-rotation axis $\bar{C}_2(z)$ (parallel to the z-axis) and an inversion at the center i of every C—C bond. These three symmetry operations together with the identity operation E form the factor group of the line group C_{2h}. The character table of this factor group is given in Table XV.

A rigorous vibrational analysis could only be based on a three-dimensional space group. But since the crystal structure of the polymer is not known, the line group treatment is a good approximation.

TABLE XV

Character Table for Line Group C_{2h} for Polyhydroxymethylene

	Symmetry elements				Number of normal vibrations			
Species	E	$\bar{C}_2(z)$	$\sigma_h(xy)$	i	Skeleton	Whole molecule	Nongenuine vibrations	IR activity
A_g	$+1$	$+1$	$+1$	$+1$	4	8	R_z	—
A_u	$+1$	$+1$	-1	-1	2	4	T_z	M_z
B_g	$+1$	-1	-1	$+1$	2	4	—	—
B_u	$+1$	-1	$+1$	-1	4	8	$T_x T_y$	$M_x M_y$

Normal Vibrations and Selection Rules

The translational repeat unit, consisting of two CHOH groups, contains 8 atoms; therefore there are 24 factor group normal vibrations.

It is convenient to consider first vibrations that mainly involve stretching and deformation of carbon-carbon and carbon-oxygen bonds. The carbon-oxygen skeleton has four atoms per repeat unit and therefore only 12 normal vibrations. They are distributed among the four species as shown in column 6 of Table XV. The geometrical form of these vibrations is shown in Fig. 17. Four of the vibrations are nongenuine ones as marked in column 8, where the conventional nomenclature is used for translations (T) and rotations (R). The selection rules for infrared spectra are listed in column 9 where M_x, M_y, and M_z give the direction of the transition moment for the particular species. Only vibrations of species A_u and B_u are infrared-active. In a stretched sample where the z-axis of the chain is predominantly parallel to the stretching direction the A_u vibrations lead to parallel bands while the B_u vibrations result in perpendicular bands if polarized infrared radiation is used.

If we include the hydrogen atoms in the vibrational anlysis we find that there are a total of eight vibrations of species A_g and of species B_u and four vibrations of species A_u and B_g, respectively (see column 7, Table XV). The 12 vibrations of species A_u and B_u are expected to be infrared active; three are pure translations, the remaining nine are shown in Fig. 18. They are numbered arbitrarily ν_1 to ν_9. The assignment of these vibrations to observed absorption bands is discussed in the following section.

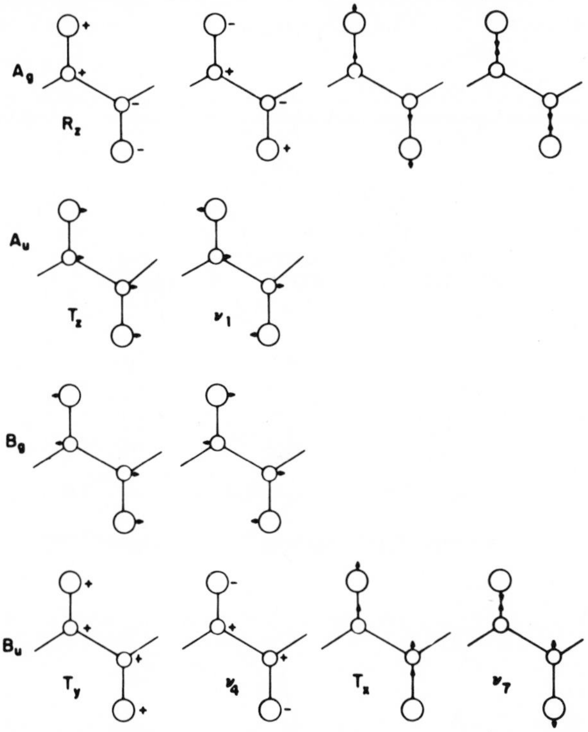

FIG. 17. Normal vibrations of the carbon-oxygen skeleton of polyhydroxymethylene Oxygen atom ◯; Carbon atom ○.

Observed Spectra and Band Assignments

Figure 19a shows polarization spectra of a film strip of polyhydroxymethylene, drawn four times. The spectrum of another less oriented film that had been soaked in D_2O is shown in Fig. 19b. From the relative intensity of the OH and OD stretching vibrations the degree of deuteration of the hydroxyl group is estimated to be of the order of 85%. The spectrum shown in Fig. 19b had to be recorded with the sample kept under nitrogen, otherwise the deuterium would exchange quite rapidly with hydrogen of atmospheric water vapor.

Based on the molecular model described in the previous sections an attempt is made to assign some of the nine infrared active normal vibrations to the observed absorption bands. The results are given in Table XVI. Both, deuteration and polarization spectra help considerably in the assignment of bands. There is little doubt about the assignment of the vibrations ν_8 and ν_9 to the OH-stretching and CH-stretching vibrations, respectively. The strong perpendicular band at 1410 cm^{-1} disappears on deuteration; it is shifted at least to about 1150 cm^{-1} or lower. It therefore can be identified with ν_5. In (CHOD)$_x$ two bands are observed between 1200 and 1400 cm^{-1}. Since this is the region for CH-bending and -wagging

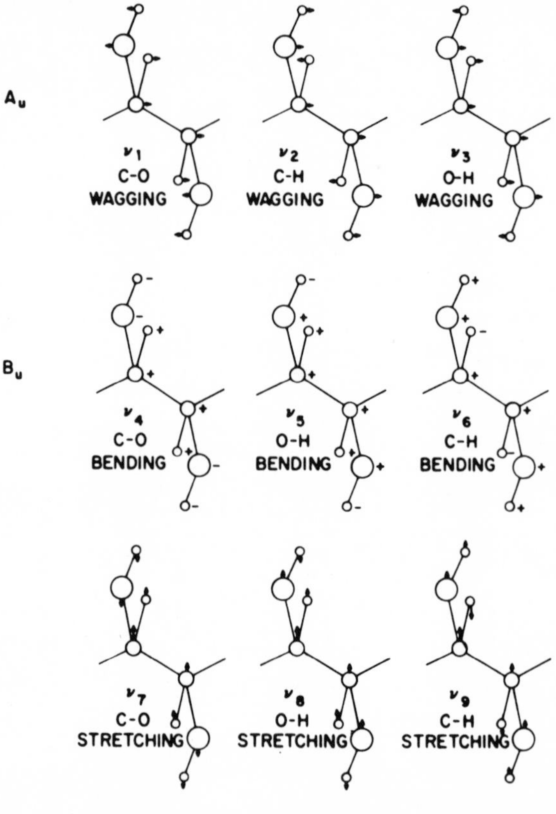

FIG. 18. Infrared active normal vibrations of polyhydroxymethylene. Oxygen atom ◯; carbon atom ◯; Hydrogen atom ○.

Footnote to Fig. 18

The vibrations of species B_u shown in this figure correspond approximately only to the designation given, for example, ν_6 and ν_9 are not pure CH-bending and -stretching vibrations, respectively. A pure CH-stretching vibration would correspond to a combination of ν_6 and ν_9. Since both these vibrations belong to the same species two such combinations corresponding to the pure CH-bending and CH-stretching vibration could also represent two basic normal vibrations instead of the ones shown in this figure. In analogy combinations of ν_4 and ν_7 as well as ν_5 and ν_8 would be pure CO and OH bending and stretching vibrations, respectively.

vibrations in other compounds such as polyethylene or polyvinylalcohol (Ref. 17) the 1350 cm^{-1} ($\perp$?) vibration can be assigned to ν_6 and the 1240 cm^{-1} ($\parallel$) band to ν_2, based on the observed dichroism. These two bands cannot be observed clearly

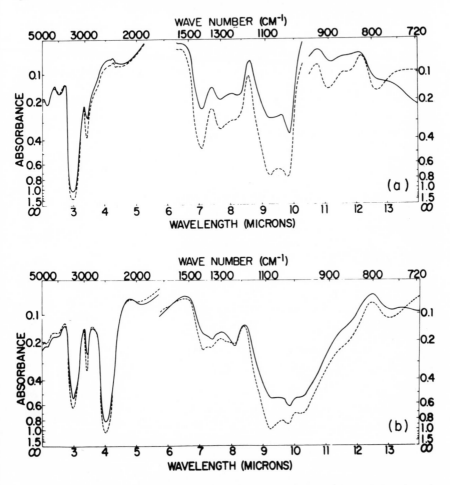

FIG. 19. Polarized infrared spectra of (a) oriented –(–CHOH–)–$_n$ films and (b) oriented –(–CHOD–)–$_n$ films from 5000 to 720 cm^{-1}. KEY: electric vector perpendicular to direction of stretch, – – –; electric vector parallel to direction of stretch, ———.

in (CHOH)$_x$ partially because of interference with the strong OH-bending vibration band. The CO-stretching vibration ν_7 is expected to absorb at about 1100 cm^{-1} in analogy to polyvinylalcohol (Ref. 17). Two bands are actually observed at 1082 and 1018 cm^{-1} for (CHOH)$_x$ and a third band appears at about 983 cm^{-1} in (CHOD)$_x$. A parallel band is found at 620 cm^{-1} (Ref. 28). It shifts below 500 cm^{-1} on deuteration and is therefore assigned to the OH-wagging vibration ν_3. ν_1, and ν_4 are expected to absorb below 500 cm^{-1}.

The assignment of absorption bands given in Table XVI is only a first crude attempt to a complete analysis. It seems that the simple molecular model described above cannot account for a number of bands such as the two perpendicular bands at 790 and 900 cm^{-1} for (CHOH)$_x$ and at 760 and 860 cm^{-1} for (CHOD)$_x$. The spectrum of (CHOH)$_x$ also shows additional bands at about 1215 and 1300 cm^{-1}. Another feature that cannot be accounted for by the simple model is the fact that

TABLE XVI

Vibration (Fig. 18)	Species	Group vibration	Observed frequency (cm^{-1})		Some frequencies[a] of corresponding group vibrations for:	
			(CHOH)$_x$	(CHOD)$_x$	(CH$_2$CHOH)$_x$	(CH$_2$CHOD)$_x$
ν_8	B_u	OH stretching	3370	2475	3340	2480
ν_9	B_u	CH stretching	2930	2930	2840	
ν_5	B_u	OH bending	1410	?	1376	
					1326 }[b]	
ν_6	B_u	CH bending	?	1350	1320	1360
ν_2	A_u	CH wagging	?	1240	1235	
ν_7	B_u	CO stretching	1082	1082		
			1018	1018	1096	
				983		
ν_3	A_u	OH wagging	620	< 500	610	
					630	
ν_4	B_u	CO bending	—	—	480	
ν_1	A_u	CO wagging	—	—	410	

[a] After Krimm (Ref. 17).
[b] Combination of CH- and OH-bending vibrations (Ref. 17).

two bands are observed in the region 1000–1100 cm^{-1} where ν_7 is expected to absorb.

To analyze the spectrum more completely additional information is required. It will be important to know the crystal structure of this polymer to carry out a vibrational analysis for the space group unit cell. This may lead to somewhat different selection rules. If there is more than one polymer chain passing through the unit cell one would expect to observe band splittings due to intermolecular interaction. It will also be important to study polymer samples with different degrees of crystallinity to distinguish between absorption bands of the crystalline and the amorphous phase. Since the amorphous polymer chains have no symmetry, all vibrations are expected to be infrared active. It will furthermore be useful to observe spectra at frequencies below 500 cm^{-1} to find the absorption bands for the vibrations ν_1 and ν_4.

REFERENCES

1. Bethe, H. A., *Ann. Physik* [5] **3**, 133 (1929).
2. Bhagavantam, S., and Venkatarayudu, T., *Proc. Indian Acad. Sci.* **A9**, 224 (1939).
3. Bhagavantam, S., and Venkatarayudu, T., "The Theory of Groups and Its Application to Physical Problems," 2nd ed. Andhra Univ. Press, Waltair, India, 1951.
4. Born, M., *Proc. Phys. Soc. (London)* **54**, 362 (1942).
5. Buerger, M. J., "Elementary Crystallography." Wiley, New York, 1956.
6. Bunn, C. W., *Trans. Faraday Soc.* **35**, 483 (1939).
7. Davydov, A. S., *Zh. Eksperim.; Teor. Fiz.* **18**, 210 (1948).
8. Davydov, A. S., "Theory of Light Absorption of Molecular Crystals." Inst. Phys. Ukrainian Acad. Sci., Kiev, 1951.
9. Field, N. D., and Schaefgen, J. R., *J. Polymer. Sci.* **58**, 533 (1962).
10. Halford, R. S., *J. Chem. Phys.* **14**, 8 (1946).
11. Herzberg, G., "Infrared and Raman Spectra of Polyatomic Molecules." Van Nostrand, Princeton, New Jersey, 1945.
12. Higgs, P. W., *Proc. Roy. Soc.* **A220**, 472 (1953).
13. Hornig, D. F., *J. Chem. Phys.* **16**, 1063 (1948).
14. Huggins, M. L., *J. Chem. Phys.* **13**, 37 (1945).
15. King, W., Hainer, R. M., and McMahon, H. O., *J. Appl. Phys.* **20**, 559 (1949).
16. Krimm, S., *J. Chem. Phys.* **22**, 567 (1954).
17. Krimm, S., *Fortschr. Hochpolymer. Forsch.* **2**, 51 (1960).
18. Krimm, S., Liang, C. Y., and Sutherland, G. B. B. M., *J. Chem. Phys.* **25**, 549 (1956).
19. Liang, C. Y., *J. Mol. Spectry.* **1**, 61 (1957).
20. Müller, A., *Proc. Roy. Soc.* **A120**, 437 (1928).
21. Nichols, J. B., *J. Appl. Phys.* **25**, 840 (1954).
22. Nielsen, J. R., and Holland, R. F., *J. Mol. Spectry.* **4**, 488 (1960).
23. Nielsen, J. R., and Woollett, A. H., *J. Chem. Phys.* **26**, 1391 (1957).
24. Novak, I. I., *Bull. Acad. Sci. U.S.S.R., Phys. Ser. (English Transl.)* **22**, 1103 (1958).
25. Raghavacharyulu, I. V. V., *Can. J. Phys.* **39**, 830 (1961).
26. Raghavacharyulu, I. V. V., and Shrestha, C. B., *J. Mol. Spectry.* **7**, 46 (1961).
27. Rosenthal, J. E., and Murphy, G. M., *Rev. Mod. Phys.* **8**, 317 (1936).
28. Schaefgen, J. R., and Zbinden, R., *J. Polymer. Sci.* (to be published).
29. Sheppard, N., *Advan. Spectry.* **1**, 288 (1959).
30. Snyder, R. G., *J. Chem. Phys.* **27**, 969 (1957).
31. Snyder, R. G., *J. Mol. Spectry.* **4**, 411 (1960).
32. Speiser, A., "Die Theorie der Gruppen von endlicher Ordnung." Dover, New York, 1945 (1st ed., Springer, Berlin, 1937).
33. Sutherland, G. B. B. M., and Sheppard, N., *Nature* **159**, 739 (1947).
34. Tadokoro, H., Yasumoto, T., Murahashi, S., and Nitta, I., *J. Polymer. Sci.* **44**, 266 (1960).
35. Tobin, M. C., *J. Chem. Phys.* **23**, 891 (1955).
36. Wigner, E. P., "Group Theory." Academic Press, New York, 1959.
37. Winston, H., and Halford, R. S., *J. Chem. Phys.* **17**, 607 (1949).
38. Woollett, A. H., Ph.D. Thesis, University of Oklahoma, 1956.

—III—
Numerical Calculations of Vibrations in Chain Molecules

1. Introduction and Historical Background

Our task in this chapter is to calculate the vibrational frequencies of certain models of chain molecules in order to learn more about the skeletal vibrations of actual chains.

High polymer molecules can be considered, in a first approximation, as chains of point masses connected by chemical bonds with certain stretching-, bending- and twisting-force constants. For some models it is possible to set up the Newtonian equations of motion and calculate the normal vibrations.

This problem was solved, at least in principle, about 200 years ago when John and Daniel Bernoulli calculated the vibrational frequencies for a one-dimensional lattice of point masses. A somewhat related problem was solved by Euler who gave a mathematical treatment of the vibrations in a continuous string, while Lagrange showed that the continuous string is the limiting case of a linear chain of point masses (Ref. 10). Born and von Kármán (Ref. 3) treated a linear monatomic, a linear diatomic and a three-dimensional diatomic lattice (e.g., NaCl) and calculated normal vibrations and the specific heat for these models.

Further information on the historical background can be found in Brillouin's book (Ref. 4).

In Section 2 we will give an account of the longitudinal vibrations of a linear monatomic lattice. The general method will be outlined for calculating the frequency as well as the geometrical form of each normal vibration. The boundary conditions for finite chain lengths are discussed in detail.

In Sections 3–5 we will consider longitudinal and transverse vibrations for linear diatomic and polyatomic lattices and discuss in detail a planar zig-zag chain for which polyethylene and some vinyl- and vinylidene-polymers are practical examples.

The methods described in this chapter may also be applied to other, more complicated chains, but the calculations might become very tedious.

2. The Classical Example of Longitudinal Vibrations in a Monatomic Linear Lattice

A. DESCRIPTION OF MODEL

Let us consider a row of N point masses m, mechanically connected by springs with a stretching force constant α (Fig. 1). The masses are located along the ξ-axis with mass number 1 at $\xi = 0$. The coordinate ξ_n of the nth mass in its equilibrium position is $(n-1)d$, where d is the distance between two adjacent masses. This is a model where only nearest-neighbor interactions are considered. A rigorous treatment, including interactions between all point masses, is given by Brillouin (Ref. 4, p. 26).

The Newtonian equations for the longitudinal motions of the point masses will be solved for the free and the fixed boundary chain.

In free boundary chains the end groups are identical with the repeat unit (Fig. 1a) while for fixed boundary chains the end groups are connected to walls (Fig. 1b). The two models are extreme cases. A finite size single crystal is an example for a free end chain whereas polymer molecules with heavy end groups or chain sections of block copolymers are examples close to the fixed boundary model.

FIG. 1. Model for a monatomic linear lattice of N point masses.

B. Equations of Motion

The equations of motion for a free and a fixed boundary chain are (the two cases will be discussed separately on the left and right sides of the following pages):

Free boundary (Fig. 1a)

Fixed boundary (Fig. 1b)

$$m\ddot{x}_1 = -\alpha(x_1 - x_2)$$
$$\vdots$$
$$m\ddot{x}_n = -\alpha(2x_n - x_{n-1} - x_{n+1})$$
$$\vdots$$
$$m\ddot{x}_N = -\alpha(x_N - x_{N-1})$$

$$m\ddot{x}_1 = -\alpha(2x_1 - x_2)$$
$$\vdots$$
$$m\ddot{x}_n = -\alpha(2x_n - x_{n-1} - x_{n+1})$$
$$\vdots$$
$$m\ddot{x}_N = -\alpha(2x_N - x_{N-1})$$

(1a, b)

x_n is the dislocation of the nth mass from its equilibrium position.

C. Calculation of Vibrational Frequencies

The frequencies as well as the mass dislocations for all possible normal vibrations will be calculated for the fixed and free boundary system assuming a solution periodic in time with a circular frequency ω:

$$x_n = X_n e^{i\omega t} \tag{2}$$

where X_n is the maximum dislocation of the nth atom from its equilibrium position.

With this solution (1a) and (1b) become systems of N homogeneous linear equations:

$$x_1\left(\frac{\omega^2}{\omega_0^2}-1\right)+x_2 \qquad = 0 \qquad\qquad x_1\left(\frac{\omega^2}{\omega_0^2}-2\right)+x_2 \qquad = 0$$

$$\vdots \qquad\qquad\qquad\qquad\qquad\qquad \vdots$$

$$\ldots x_{n-1}+x_n\left(\frac{\omega^2}{\omega_0^2}-2\right)+x_{n+1} = 0 \qquad \ldots x_{n-1}+x_n\left(\frac{\omega^2}{\omega_0^2}-2\right)+x_{n+1} = 0 \quad \text{(3a, b)}$$

$$\vdots \qquad\qquad\qquad\qquad\qquad\qquad \vdots$$

$$\ldots\ldots x_{N-1}+x_N\left(\frac{\omega^2}{\omega_0^2}-1\right) \quad = 0 \qquad\qquad \ldots\ldots x_{N-1}+x_N\left(\frac{\omega^2}{\omega_0^2}-2\right) \quad = 0$$

where $\omega_0^2 = m/\alpha$. Equations (3a) and (3b) have a solution only if the respective secular determinants are equal to zero:

$$
\begin{vmatrix}
\frac{\omega^2}{\omega_0^2}-1, & 1, & 0, & \cdot & \cdot & \cdot & \cdot & \cdot & \cdot & 0 \\
1, & \frac{\omega^2}{\omega_0^2}-2, & 1, & 0, & \cdot & \cdot & \cdot & \cdot & \cdot & 0 \\
0, & 1, & \frac{\omega^2}{\omega_0^2}-2, & 1, & 0, & \cdot & \cdot & \cdot & \cdot & 0 \\
\vdots & & & & & & & & & \vdots \\
0, & \cdot & & \cdot & \cdot & \cdot & 0, & 1, & \frac{\omega^2}{\omega_0^2}-2, & 1 \\
0, & \cdot & & \cdot & \cdot & \cdot & 0, & & 1, & \frac{\omega^2}{\omega_0^2}-1
\end{vmatrix}_N = 0 \quad \text{(4a)}
$$

$$
\begin{vmatrix}
\frac{\omega^2}{\omega_0^2}-2, & 1, & 0, & \cdot & \cdot & \cdot & \cdot & \cdot & \cdot & 0 \\
1, & \frac{\omega^2}{\omega_0^2}-2, & 1, & 0, & \cdot & \cdot & \cdot & \cdot & \cdot & 0 \\
0, & 1, & \frac{\omega^2}{\omega_0^2}-2, & 1, & 0, & \cdot & \cdot & \cdot & \cdot & 0 \\
\vdots & & & & & & & & & \vdots \\
0, & \cdot & & \cdot & \cdot & \cdot & 0, & 1, & \frac{\omega^2}{\omega_0^2}-2, & 1 \\
0, & \cdot & & \cdot & \cdot & \cdot & 0, & & 1, & \frac{\omega^2}{\omega_0^2}-2
\end{vmatrix}_N = 0 \quad \text{(4b)}
$$

These secular equations of the order N can be solved for ω^2. A detailed mathematical treatment is given in a book by Houston (Ref. 7) and in two papers by Rutherford (Refs. 17, 18). The N roots ω_s^2 of (4a) and (4b) are

$$\omega_s^2 = 4\omega_0^2 \cos^2 \frac{s\pi}{2N} \qquad \omega_s^2 = 4\omega_0^2 \cos^2 \frac{s\pi}{2(N+1)}$$

$$= 4\omega_0^2 \sin^2 \frac{(N-s)\pi}{2N} \qquad = 4\omega_0^2 \sin^2 \frac{(N+1-s)\pi}{2(N+1)} \qquad \text{(5a, b)}$$

where $s = 1, 2, \ldots, N$. Equation (5b) is exactly the solution given by Lagrange (Ref. 10) for the N normal vibration frequencies. The values of ω_s fall into the interval ranging from 0 to $2\omega_0$. The distribution of the frequencies in this interval will be discussed in Section 2G.

D. Geometrical Form of the Normal Vibrations

We now will calculate the dislocation of each point mass during a vibration. Let us consider one specific normal vibration with the frequency ω_s. Substituting for ω in Eq. (3) the values ω_s of Eq. (5) we obtain a system of N homogeneous equations in $x_1, x_2, \ldots, x_N$:

$$x_1\left(4\cos^2\frac{s\pi}{2N} - 1\right) + x_2 = 0 \qquad x_1\left(4\cos^2\frac{s\pi}{2(N+1)} - 2\right) + x_2 = 0$$

$$\vdots \qquad\qquad\qquad \vdots$$

$$\ldots x_{n-1} + x_n\left(4\cos^2\frac{s\pi}{2N} - 2\right) \qquad \ldots x_{n-1} + x_n\left(4\cos^2\frac{s\pi}{2(N+1)} - 2\right)$$

$$+ x_{n+1} = 0 \qquad\qquad\qquad + x_3 = 0 \quad \text{(6a, b)}$$

$$\vdots \qquad\qquad\qquad \vdots$$

$$\ldots\ldots x_{N-1} \qquad\qquad \ldots\ldots x_{N-1}$$

$$+ x_N\left(4\cos^2\frac{s\pi}{2N} - 2\right) = 0 \qquad + x_N\left(4\cos^2\frac{s\pi}{2(N+1)} - 2\right) = 0$$

These equations have N solutions which are determined within a constant factor. They can be found by the standard method for solving systems of linear equations (Refs. 17, 18), and expressed in the following form:

$$X_n^s = \text{const}\,(-1)^n \sin\left(\frac{sn}{N} - \frac{s}{2N}\right)\pi \qquad X_n^s = \text{const}\,(-1)^n \sin\frac{sn}{N+1}\pi \quad \text{(7a, b)}$$

where X_n^s is the maximum displacement of the nth mass for a normal vibration of the frequency ω_s.

E. ILLUSTRATION

Equation (7) describes sinusoidal standing waves along the linear chain of N point masses. In order to show this more clearly the X_n^s values have been calculated for the specific case of eight masses. Figures 2a and 2b show the actual displacements for the eight vibrations $\omega_1 \ldots \omega_8$. The reader should remember that we

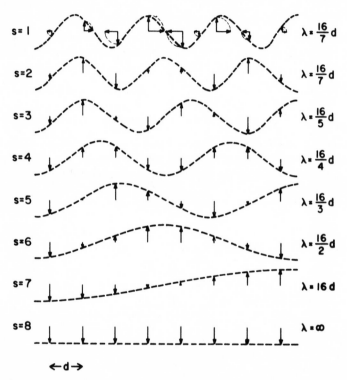

FIG. 2a. Longitudinal vibrations of a linear chain of 8 point masses with free ends. The real dislocations of the masses are obtained by turning the vectors by 90° clockwise as indicated for $s = 1$.

consider here only longitudinal vibrations (one degree of freedom for one mass), but for convenience the dislocations of the masses in Figs. 2a and 2b are drawn in a direction perpendicular to the chain, so that the sinusoidal standing wave can easier be visualized. The actual dislocation is obtained by turning the vectors drawn in Figs. 2a and 2b by 90° clockwise as indicated for the value $s = 1$.

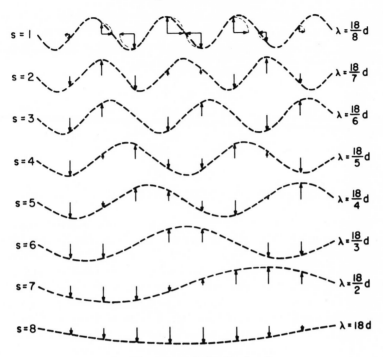

$s = 1$ $\lambda = \frac{18}{8}d$

$s = 2$ $\lambda = \frac{18}{7}d$

$s = 3$ $\lambda = \frac{18}{6}d$

$s = 4$ $\lambda = \frac{18}{5}d$

$s = 5$ $\lambda = \frac{18}{4}d$

$s = 6$ $\lambda = \frac{18}{3}d$

$s = 7$ $\lambda = \frac{18}{2}d$

$s = 8$ $\lambda = 18d$

FIG. 2b. Same as 2a but fixed ends.

F. Wavelength† of Standing Waves and Phase Shift between Two Point Masses

In this section we first will briefly describe the geometrical form of the standing waves by determining their wavelengths.

The wavelength λ_s of a standing wave ω_s in a one-dimensional chain of point masses with free and fixed ends is given by the expressions

$$\lambda_s = \frac{2N}{N-s}d \quad \bigg| \quad \lambda_s = \frac{2(N+1)}{N+1-s}d \qquad \text{(8a, b)}$$

Equations (8a, b) can easily be verified for a specific example with the help of Figs. 2a and 2b.

At this point we introduce the wave number $\mathbf{k}_s$ which is defined by

$$\mathbf{k}_s = \frac{2\pi}{\lambda_s} \qquad \text{(9)}$$

† This is *not* a wavelength of light.

where $s = 1, \ldots, N$ and discuss the more direct way of calculating vibrational frequencies in a chain. This method was originally used by Lagrange (Ref. 10) and later by Born and von Kármán (Ref. 3). We make use of Eqs. (8a, b) and (9) and write the solution (7a, b) in the following form:

$$X_n^s = \text{const} \sin\left(-\mathbf{k}_s nd + \varphi_s\right) \tag{10}$$

where the phase constant $\varphi_s = 0$ for a fixed boundary and $\varphi_s = s\pi/2N$ for a free boundary chain. A vibration is, of course, also periodic in time with the frequency ω_s. Including this time dependence and expressing Eq. (10) in the conventional complex form we obtain

$$x_n^s = A_s \exp\left[i(\omega_s t - \mathbf{k}_s nd + \varphi_s)\right] \tag{11}$$

where A_s is the amplitude of the vibration. Such a solution, periodic in space and time, was assumed by Born and von Kármán (Ref. 3) in order to solve Eq. (1b) (fixed boundary) directly. This direct method is simpler from a mathematical point of view and will be used in further calculations on diatomic and zig-zag chains, but the method presented in this section was chosen to calculate explicitly the value of the wavelength λ_s (or $\mathbf{k}_s$, respectively) and the phase shift ϕ_s between two successive point masses for two boundary conditions. This point caused some controversy in the literature (Refs. 8, 9, 16), and we therefore express ϕ_s for free and fixed boundary chains in various forms. From (11) we can easily see that in a standing wave the phase shift between the point mass n and point mass $n+1$ is given by

$$\phi_s = \mathbf{k}_s d = \frac{2\pi d}{\lambda_s} \tag{12}$$

We substitute λ_s by the values of (8a, b) and obtain for ϕ_s

(Free boundary)   (Fixed boundary)

$$\phi_s = \frac{\pi(N-s)}{N} \qquad\qquad \phi_s = \frac{\pi(N+1-s)}{N+1} \tag{13}$$

By changing the running number s in the following way:

$$j = N - s \qquad\qquad i = N + 1 - s$$

Eq. (13) assumes the form

$$\phi_s = \frac{\pi j}{N} \qquad\qquad \phi_s = \frac{\pi i}{N+1} \tag{14}$$

$\text{where } j = 0, 1, \ldots, N-1 \qquad\qquad \text{where } i = 1, 2, \ldots, N$

G. Frequency Branch

Finally, it is interesting to notice that the expressions for the frequencies (5a) and (5b) are the same for free and fixed boundary chains if $\mathbf{k}_s$ or λ_s is introduced as a variable:

$$\omega_s = 2\omega_0 \sin \frac{\mathbf{k}_s d}{2} = 2\omega_0 \sin \frac{\pi d}{\lambda_s} \tag{15}$$

This equation shows that the frequency ω_s is a periodic function of the wave number $\mathbf{k}_s$. It is conventional to plot this relationship as a continuous function, shown in Fig. 3, where the form of the curve is independent of the chain length.

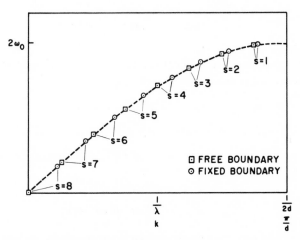

FIG. 3. Frequency branches for the longitudinal vibrations in a chain of 8 point masses with fixed and free boundaries. Plotted is the frequency against the wave number k as well as $1/\lambda$.

This curve is called a frequency branch. Figure 3 also shows eight discrete frequencies for a chain of eight point masses with a free and a fixed boundary, respectively. A longer chain would have more points on the same branch and for an infinitely long chain the infinite number of frequencies would cover continuously the interval from 0 to $2\omega_0$.

3. Diatomic One-Dimensional Lattice

A. Longitudinal Vibrations

In this section we will present the calculation of the longitudinal vibrations for a diatomic linear lattice. A one-dimensional sodium chloride " crystal " would be a good example for such a chain. m_1 and m_2 are the two kinds of masses, α_1 and α_2 are the force constants, d_1 and d_2 the distances as shown in Fig. 4.

We will not distinguish between a fixed and a free end chain (Ref. 24) since we are going to assume a general solution of the form of Eq. (11). The frequencies will be calculated but not the actual displacements of the masses, since in this section we only want to explain the method, rather than to give detailed calculations, which can be found in textbooks on solid state physics (see, e.g., Ref. 5) or in Brillouin's book (Ref. 4).

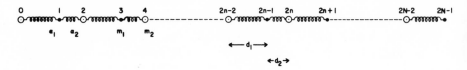

FIG. 4. Diatomic one-dimensional lattice.

The equations of motion consist of two sets for the two masses m_1 and m_2. Only one representative equation for each set is given:

$$m_2 \ddot{x}_{2n} = -\alpha_2(x_{2n}-x_{2n-1})-\alpha_1(x_{2n}-x_{2n+1})$$
$$m_1 \ddot{x}_{2n+1} = -\alpha_1(x_{2n+1}-x_{2n})-\alpha_2(x_{2n+1}-x_{2n+2}) \tag{16}$$

A solution of the form of (11) is assumed where the constant phase shift φ is neglected:

$$x_{2n} = A_2 \exp[i(\omega t - \mathbf{k}nd)]$$
$$x_{2n+1} = A_1 \exp[i(\omega t - \mathbf{k}(nd+d_1))] \tag{17}$$

A_1 and A_2 are the maximum displacements of the masses m_1 and m_2. From (16) and (17) we obtain

$$(m_2\omega^2 - \alpha_1 - \alpha_2)x_{2n} + (\alpha_1 + \alpha_2 e^{ikd})x_{2n+1} = 0$$
$$(\alpha_1 + \alpha_2 e^{-ikd})x_{2n} + (m_1\omega^2 - \alpha_1 - \alpha_2)x_{2n+1} = 0 \tag{18}$$

This system of two homogeneous equations has a solution only if its secular determinant disappears. The quadratic equation resulting from this condition can be solved for ω^2 to give the two roots:

$$\omega_{\pm}^2 = \frac{\alpha_1+\alpha_2}{2}\left(\frac{1}{m_1}+\frac{1}{m_2}\right)\pm\sqrt{\left(\frac{\alpha_1+\alpha_2}{2}\right)^2\left(\frac{1}{m_1}+\frac{1}{m_2}\right)^2 - 4\frac{\alpha_1\alpha_2}{m_1m_2}\sin^2\frac{d\mathbf{k}}{2}} \tag{19}$$

Let us now discuss this result.

Frequency Branches

The plus and minus signs have a physical meaning. They lead to two sets of N solutions since the wave number $\mathbf{k}$ can assume the N discrete values $\mathbf{k}_s$ given by

Eqs. (8a, b) and (9). The two sets fall on two frequency branches. With the plus sign we obtain the optical branch, with the minus sign the acoustical branch as shown in Figs. 5a and 5b.

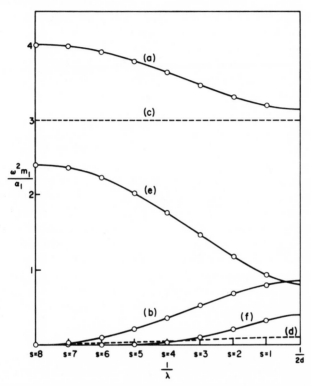

FIG. 5. Frequency branches for a diatomic one-dimensional lattice where $m_1 = 2m_2$. ○ indicate frequencies for a lattice with free ends, consisting of 8 diatomic molecules. *Longitudinal vibrations*: (a) optical branch $\alpha_1 = 3\alpha_2$; (b) acoustical branch $\alpha_1 = 3\alpha_2$; (c) optical branch, loose coupling, $\alpha_2 \ll \alpha_1$; (d) acoustical branch, loose coupling, $\alpha_2 = \alpha_1/25$. *Transverse vibrations*: (e) optical branch, doubly degenerate, for $\beta = \alpha_1 d/20$; (f) acoustical branch, doubly degenerate, for $\beta = \alpha_1 d/20$.

Loose Coupling of Diatomic Molecules in a Chain

The case where the two force constants are highly different, for example, $\alpha_1 \gg \alpha_2$, corresponds to a chain of diatomic molecules with very loose coupling between the molecules. Equation (19) can then be approximated by

$$\omega_+^2 \approx \frac{\alpha_1}{\mu} \tag{20a}$$

$$\omega_-^2 \approx \frac{\alpha_2}{m_1 + m_2} 4\sin^2\frac{d\mathbf{k}}{2} \tag{20b}$$

where μ is the reduced mass defined by

$$\mu = \frac{m_1 m_2}{m_1 + m_2} \tag{21}$$

Equation (20a) gives the vibrational frequency of a free diatomic molecule and is independent of $\mathbf{k}$ which means that the optical branch reduces to one frequency only. Equation (20b) is the same as (15) where α corresponds to α_2 and the mass m to $m_1 + m_2$. This result is obvious if each diatomic molecule is replaced by a point mass m. The two branches corresponding to (20) are represented by dotted lines in Fig. 5.

B. TRANSVERSE VIBRATIONS

So far we have considered only longitudinal vibrations, but if we introduce bending force constants for the chain and allow perpendicular motions we have a total of three degrees of freedom (one longitudinal and two transverse) for each point mass in the chain. The Newtonian equations of motion in the y- and z- directions (perpendicular to the chain) are slightly more complicated than (16), but the form of the solution for the frequencies shows again the existence of an optical and an acoustical branch. Both are doubly degenerate since the bending force constants are assumed to be the same in the y- and z-directions.

The equations of motion for perpendicular vibrations involve the two nearest as well as the two second nearest point masses. We assume a bending force constant β and for simplicity an equal distance between the mass points: $d_1 = d_2 = d/2$. The force on a mass m_{2n} is

$$F_{2n} = \beta(\varDelta\gamma_{2n-1} - 2\varDelta\gamma_{2n} + \varDelta\gamma_{2n+1}) \tag{22}$$

where $\varDelta\gamma_i$ is the deviation from the equilibrium angle ($180°$) at the point mass m_i. By replacing $\varDelta\gamma_i$ by the actual displacements of the point masses perpendicular to the chain the Newtonian equations can be written in the following form:

$$m_2 \ddot{y}_{2n} = \frac{2\beta}{d}(-y_{2n-2} + 4y_{2n-1} - 6y_{2n} + 4y_{2n+1} - y_{2n+2}) \tag{23a}$$

$$m_1 \ddot{y}_{2n+1} = \frac{2\beta}{d}(-y_{2n-1} + 4y_{2n} - 6y_{2n+1} + 4y_{2n+2} - y_{2n+3}) \tag{23b}$$

and with a solution similar to (17) the following equation for the frequency is obtained:

$$\frac{m_1 m_2 d\omega^2}{2\beta} = (m_1 + m_2)(\cos d\mathbf{k} + 3)$$

$$\pm \sqrt{(m_1 + m_2)^2(\cos d\mathbf{k} + 3)^2 - 4m_1 m_2(\cos^2 d\mathbf{k} - 2\cos d\mathbf{k} + 1)} \tag{24}$$

Acoustical ($-$ sign) and optical ($+$ sign) branches are shown in Fig. 5 for a specific example.

C. SUMMARY

The distribution of the frequencies in a diatomic one-dimensional chain can be described in the following way. They are grouped into six frequency branches, each consisting of N vibrations. Three are acoustical branches and three are optical branches. The actual number of branches is only four since the transverse branches are doubly degenerate. Figure 5 shows the frequency branches for a chain of eight diatomic molecules with free ends. The three zero vibrations (corresponding to translations) are obviously the end points of the acoustical branches where $\mathbf{k} = 0$.

4. Chain Molecules with an *n*-Atomic Repeat Unit

The method of evaluating the number of vibrational frequency branches described in Section 3 can be extended to polymer molecules of a more general form where the repeat unit consists of n atoms either lying on the chain axis

TYPE OF CHAINS WITH N REPEAT UNITS	LONGITUDINAL BRANCHES	TRANSVERSE BRANCHES	ACOUSTICAL BRANCHES			OPTICAL BRANCHES		
			LONGI-TUDINAL	TRANS-VERSE	TOTAL	LONGI-TUDINAL	TRANS-VERSE	TOTAL
(a) o o o o o o	1	1 d.d.	1	2	3	0	0	0
(b) o● o● o● o●	2	2 d.d.	1	2	3	1	2	3
(c) o●...+ o●...+ (n)	n	n d.d.	1	2	3	n−1	2 (n−1)	3n−3
(d) o●●...+ o●●...+ (n)	n	2n	1	3	4	n−1	2(n−1)−1	3n−4

FIG. 6. Number of frequency branches in various chains. d.d. = doubly degenerate.

(Fig. 6c) or arranged in space in an arbitrary way (Fig. 6d). The total number of vibrations and their distribution into optical and acoustical frequency branches is shown in Fig. 6. Each branch contains N vibrations, where N is the number of repeat units. The frequencies of the most general polymer chain fall on 4 acoustical and $3n - 4$ optical branches (provided that the polymer chain is straight) while in the case of a three-dimensional crystal of an *n*-atomic molecule one would have only 3 acoustical but $3n-3$ optical branches. This is due to the fact that the rotation around the chain axis is a zero vibration in a single chain but not in a crystal.

5. Planar Zig-Zag Chain

A. INTRODUCTION

A relatively simple case is the one of a planar diatomic zig-zag chain. The fundamental importance of this problem for the thermodynamics and spectro-scopy of normal hydrocarbons has been recognized long ago. Kirkwood (Ref. 8) has suggested a method to calculate the in-plane normal vibrations of a carbon zig-zag skeleton chain. His calculations were extended by Pitzer (Ref. 16) to out-of-plane vibrations and applied to polyethylene by Liang, Krimm, and Sutherland (Ref. 9).

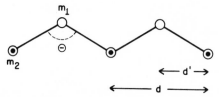

FIG. 7. Planar zig-zag chain of a vinyl polymer. m_1 = mass of CH_2 group; m_2 = mass of CHR group.

We will go one step further and extend the Kirkwood method to a zig-zag chain with alternating masses for which a vinyl polymer of the form $-(-CH_2-CHR-)_x$ is the most important practical example. R can be any substituent. Although most of the vinyl polymers do not have planar backbone chains the planar model might still be used as a first approximation.

We will consider a chain as shown in Fig. 7 where m_1 is the mass of the CH_2 group and m_2 is the mass of the CHR group. Other zig-zag polymers, of course, may be treated the same way. In the case of polytrifluoroethylene $-(-CF_2-CFH-)_x$, for example, m_1 and m_2 would be the masses of the CFH and the CF_2 groups, respectively.

This model neglects the vibrations of functional groups such as C—F or C—H stretching vibrations which will lead to some difficulties since skeletal and group vibrations can be partially mixed, especially if the frequencies are of the same order of magnitude.

As shown in Fig. 7 the repeat unit consists of two masses each having three degrees of freedom. We, therefore, would expect to find $6N$ normal vibrations, falling into six frequency branches. For an infinitely long chain the number of repeat units N becomes infinity but we still would have the frequencies grouped into the six branches. According to Section 4 and Fig. 6, four of the branches are acoustical and two are optical. One optical and three acoustical branches correspond to transverse vibrations, while one optical branch and one acoustical branch are due to longitudinal motions of the masses m_1 and m_2. As opposed to the linear chain discussed in Section 2 no vibrational branch will be degenerate in a zig-zag chain.

We now will calculate the frequencies for the in- and out-of-plane vibrations according to the Kirkwood method.

B. Equations of Motion for the In-Plane Vibrations

The solid line in Fig. 8 shows a planar zig-zag chain in equilibrium position. The dashed line represents an arbitrary in-plane displacement at a certain moment during a vibration. The displacement of the point masses from the

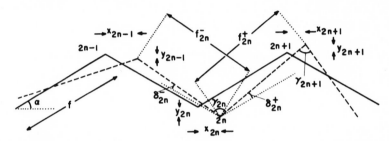

Fig. 8. Arbitrary in-plane displacement of the point masses in a planar zig-zag chain. $m_1 = m_3 = m_{2n+1}$; $m_2 = m_4 = m_{2n}$.

equilibrium position is given in Cartesian coordinates in order to set up the differential equations of motion the solution of which will lead to the actual normal vibrations of the chain.

Since we consider the in-plane vibrations only, four equations of motion will describe completely the vibrations of an infinitely long chain, because we only have two kinds of masses m_1 and m_2 each of which has two degrees of freedom. They are identified with the x- and y-directions as shown in Fig. 8. The four differential equations can be written in the following form:

$$
\begin{aligned}
m_2 \ddot{x}_{2n} = \ &k_b \sin \alpha [\sin \alpha (x_{2n-2} - 2x_{2n} - x_{2n+2}) \\
&+ \cos \alpha (-y_{2n-2} + 2y_{2n-1} - 2y_{2n+1} + y_{2n+2})] \\
&+ k_s \cos \alpha [\cos \alpha (x_{2n-1} - 2x_{2n} + x_{2n+1}) \\
&+ \sin \alpha (-y_{2n-1} + y_{2n+1})]
\end{aligned}
\tag{25a}
$$

$$
\begin{aligned}
m_1 \ddot{x}_{2n+1} = \ &k_b \sin \alpha [\sin \alpha (x_{2n-1} - 2x_{2n+1} - x_{2n+3}) \\
&- \cos \alpha (-y_{2n-1} + 2y_{2n} - 2y_{2n+2} + y_{2n+3})] \\
&+ k_s \cos \alpha [\cos \alpha (x_{2n} - 2x_{2n+1} + x_{2n+2}) \\
&- \sin \alpha (-y_{2n} + y_{2n+2})]
\end{aligned}
\tag{25b}
$$

$$m_2\ddot{y}_{2n} = k_b \cos\alpha[\sin\alpha(x_{2n-2}+2x_{2n-1}-2x_{2n+1}-x_{2n+2})$$
$$+\cos\alpha(-y_{2n-2}+4y_{2n-1}-6y_{2n}+4y_{2n+1}-y_{2n+2})]$$
$$+k_s\sin\alpha[\cos\alpha(-x_{2n-1}+x_{2n+1})$$
$$+\sin\alpha(y_{2n-1}-2y_{2n}+y_{2n+1})] \qquad (25c)$$

$$m_1\ddot{y}_{2n+1} = k_b\cos\alpha[-\sin\alpha(x_{2n-1}+2x_{2n}-2x_{2n+2}-x_{2n+3})$$
$$+\cos\alpha(-y_{2n-1}+4y_{2n}-6y_{2n+1}+4y_{2n+2}-y_{2n+2})]$$
$$+k_s\sin\alpha[-\cos\alpha(-x_{2n}+x_{2n+2})$$
$$+\sin\alpha(y_{2n}-2y_{2n+1}+y_{2n+2})] \qquad (25d)$$

k_b and k_s are the bending and stretching force constants for a C—C—C group. α (as shown in Fig. 8) is equal to $\frac{1}{2}(\pi-\gamma)$, where γ is the bond angle of the chain. It is usually close to the tetrahedral bond angle (109° 28′). The x and y values are the displacements in Cartesian coordinates of the point masses from their equilibrium position.

As an example we will indicate how Eq. (25d) was derived. The equation of motion for the y-component of atom $2n+1$ can be written in the form

$$m_1\ddot{y}_{2n+1} = F_y^{2n+1} = F_{y,b}^{2n+1}+F_{y,s}^{2n+1} \qquad (26)$$

F_y^{2n+1} is the restoring force in the y-direction on atom $2n+1$. It can be separated into a bending and a stretching term which, in turn, can be expressed as functions of the chain deformations:

$$F_{y,s}^{2n+1} = -k_s\sin\alpha(\Delta f_{2n}^{+}+\Delta f_{2n+2}^{-}) \qquad (27)$$

$$F_{y,b}^{2n+1} = k_b f\cos\alpha(\Delta\gamma_{2n}+\Delta\gamma_{2n+2}+2\Delta\gamma_{2n+1}) \qquad (28)$$

The $\Delta\gamma$ and Δf values can be expressed in terms of y and x according to Fig. 8; for example,

$$\Delta f_{2n}^{+} = (-x_{2n}+x_{2n+1})\cos\alpha+(-y_{2n}+y_{2n+1})\sin\alpha \qquad (29)$$

$$\Delta\gamma_{2n} = -\frac{1}{f}[(x_{2n-1}-x_{2n+1})\sin\alpha+(y_{2n-1}-2y_{2n}+y_{2n+1})\cos\alpha] \qquad (30)$$

Shimanouchi and Mizushima (Ref. 19) have slightly modified this simple valence force model. They introduced an additional interaction between every second point mass, that is, between m_{2n-1} and m_{2n+1} (see Fig. 8). By adjusting this additional interaction force constant as well as the bending force constant k_b they could obtain a better agreement of calculated and observed frequencies for n-butane and n-pentane than with the simpler Kirkwood method. Since we do not have enough experimental data to adjust the force constants in the case of high polymer molecules, we will proceed using Kirkwood's valence force model.

To solve Eqs. (25) we assume in analogy to (17) a solution of the form:

$$x_{2n} = A \exp\left\{i\left[\omega t + 2n\frac{d}{2}\mathbf{k}\right]\right\}, \qquad x_{2n+1} = A' \exp\left\{i\left[\omega t + (2n+1)\frac{d}{2}\mathbf{k}\right]\right\}$$

$$y_{2n} = B \exp\left\{i\left[\omega t + 2n\frac{d}{2}\mathbf{k}\right]\right\}, \qquad y_{2n+1} = B' \exp\left\{i\left[\omega t + (2n+1)\frac{d}{2}\mathbf{k}\right]\right\}$$

(31)

where d is again the length of the repeat unit (Fig. 7). This solution combined with Eqs. (25) leads to a set of four homogeneous equations in A, A', B, and B':

$$\left(m_2\omega^2 - G - 2F\sin^2\frac{d\mathbf{k}}{2}\right)A + G\cos\frac{d\mathbf{k}}{2}A' + P\sin d\mathbf{k}\,iB$$
$$+ (N-2P)\sin\frac{d\mathbf{k}}{2}iB' = 0 \quad (32\text{a})$$

$$-P\sin d\mathbf{k}\,iA + (N-2P)\sin\frac{d\mathbf{k}}{2}iA' + [m_2\omega^2 - R - Q(\cos d\mathbf{k} + 3)]B$$
$$+ (4Q+R)\cos\frac{d\mathbf{k}}{2}B' = 0 \quad (32\text{b})$$

$$G\cos\frac{d\mathbf{k}}{2}A + \left(m_1\omega^2 - G - 2F\sin^2\frac{d\mathbf{k}}{2}\right)A' + (2P-N)\sin\frac{d\mathbf{k}}{2}iB$$
$$- P\sin d\mathbf{k}\,iB' = 0 \quad (32\text{c})$$

$$(2P-N)\sin\frac{d\mathbf{k}}{2}iA + P\sin d\mathbf{k}\,iA' + (4Q+R)\cos\frac{d\mathbf{k}}{2}B$$
$$+ [m_1\omega^2 - R - Q(\cos d\mathbf{k} + 3)]B' = 0 \quad (32\text{d})$$

where for convenience the following abbreviations have been introduced:

$$2k_s\cos^2\alpha = G$$
$$2k_b\sin^2\alpha = F$$
$$2k_s\cos\alpha\sin\alpha = N$$
$$2k_b\cos\alpha\sin\alpha = P$$
$$2k_b\cos^2\alpha = Q$$
$$2k_s\sin^2\alpha = R$$

(33)

Equations (32) have a solution only if their secular determinant is equal to zero. This condition leads to an equation of the fourth degree in ω^2. A solution of this equation gives four values for ω^2. They depend on the wave number $\mathbf{k}$ which can assume N values [see Eqs. (8) and (9)] for a chain with N repeat units. Each of the four solutions is therefore N-fold, which means that the total number of in-plane vibrations is $4N$ as one would expect for a system of $2N$ point masses with two degrees of freedom. The four sets of solutions are

$$\omega_\nu(\mathbf{k}_s) \qquad \text{where} \quad \nu = 1, 2, 3, 4$$
$$s = 1, \ldots, N \tag{34}$$

$\mathbf{k}_s$ is the sth wave-number vector given by Eq. (9). Each of the four sets constitutes a frequency branch if plotted in a diagram similar to Fig. 5. Two are acoustical branches and two are optical branches. For an infinitely long chain the number of normal vibrations would be infinity but the four branches would remain exactly the same with the only difference being that the distribution of frequencies would be infinitely dense. This was explained already for the case of a linear diatomic chain.

C. POTENTIALLY INFRARED-ACTIVE IN-PLANE VIBRATIONS IN AN INFINITELY LONG ZIG-ZAG CHAIN

We have solved Eqs. (32) in the most general form only numerically as we will show in Section 5G, but it is possible to give explicit solutions for special cases which will be discussed in this and in the following section.

First we restrict the calculations to the case where the wave-number vector $\mathbf{k} = 0$. It is useful to consider this case separately since we know from the group theoretical considerations in the previous chapter that for an infinitely long chain only the vibrations for which $\mathbf{k} = 0$ are potentially infrared- or Raman-active. This is "almost true" for long chains of a finite length where in a given frequency branch the vibration for which $\mathbf{k}$ is closest to zero shows the highest potential activity.

For $\mathbf{k} = 0$ Eqs. (32) assume the following simplified form:

$$(m_2\omega^2 - G)A + GA' = 0 \tag{35a}$$

$$(m_2\omega^2 - R - 4Q)B + (4Q + R)B' = 0 \tag{35b}$$

$$GA + (m_1\omega^2 - G)A' = 0 \tag{35c}$$

$$(4Q + R)B + (m_1\omega^2 - R - 4Q)B' = 0 \tag{35d}$$

Equations (35a) and (35c) as well as (35b) and (35d) form each a system of two homogeneous equations, the secular equations of which are

$$\begin{vmatrix} (m_2\omega^2 - G) & G \\ G & (m_1\omega^2 - G) \end{vmatrix} = 0 \tag{36a}$$

$$\begin{vmatrix} (m_2\omega^2 - R - 4Q) & (4Q + R) \\ (4Q + R) & (m_1\omega^2 - R - 4Q) \end{vmatrix} = 0 \tag{36b}$$

The four solutions of the two quadratic equations in ω^2, (36a) and (36b), are given by

$$\omega_1^2(0) = 0 \tag{37a}$$

$$\omega_2^2(0) = \frac{G}{M} = \frac{2k_s}{M}\cos^2\alpha \tag{37b}$$

$$\omega_3^2(0) = 0 \tag{37c}$$

$$\omega_4^2(0) = \frac{R+4Q}{M} = \frac{2}{M}(k_s\sin^2\alpha + 4k_b\cos^2\alpha) \tag{37d}$$

where M is the reduced mass defined by

$$M = \frac{m_1 m_2}{m_1 + m_2} \tag{38}$$

$\omega_1(0)$ and $\omega_3(0)$ are the two end points of the acoustical branches. They both are zero vibrations and correspond to translations in the x- and y-directions. $\omega_2(0)$ and $\omega_4(0)$ are the end points of the optical branches and correspond to potentially infrared- and Raman-active vibrations.

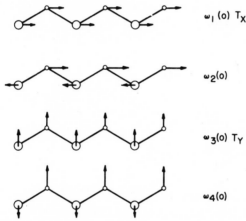

$\omega_1(0)\ T_X$

$\omega_2(0)$

$\omega_3(0)\ T_Y$

$\omega_4(0)$

FIG. 9. Potentially infrared- or Raman-active normal vibrations of an infinitely long planar zig-zag chain. T_x and T_y are translations in the x- and y-directions, respectively. $\bigcirc = m_1$ $\odot = m_2$; $m_2 = 2m_1$ in this figure.

The displacements of the masses during the four vibrations is shown in Fig. 9. They are determined by the vibrational amplitudes A, A', B, and B' which can be calculated by solving Eqs. (35) with the frequency ω substituted by the eigenvalues (37). The result of the calculation is

for the two nonzero vibrations: $\dfrac{A}{A'} = \dfrac{B}{B'} = -\dfrac{m_1}{m_2}$ (39)

for the two translations: $\dfrac{A}{A'} = \dfrac{B}{B'} = 1$ (40)

Both of these relations are as expected. Equation (39) means that the center (axis) of mass does not move during the vibration and (40) is the condition for a translation. $\omega_2(0)$ and $\omega_4(0)$ are infrared-active since the dipole moment changes during these vibrations. $\omega_2(0)$ is a parallel band and $\omega_4(0)$ is a perpendicular band. This can be observed with polarized infrared radiation for an oriented polymer sample.

For polyethylene where $m_1 = m_2 = m$ the amplitude of both masses (in Fig. 9) is the same. No change of dipole moment occurs during the vibrations $\omega_2(0)$ and $\omega_4(0)$. The in-plane skeleton vibrations of an infinitely long polyethylene chain are, therefore, expected to be infrared-inactive as we have seen already in the previous chapter.

D. In-Plane Vibrational Frequencies for a Chain with Equal Masses

If $m_1 = m_2 = m$ as in normal hydrocarbons or in polytetrafluoroethylene, Eqs. (32) can easily be solved for the most general case where the wave-number vector $\mathbf{k} \neq 0$.

Kirkwood (Ref. 8) has derived a solution for this case in a somewhat simpler way which we now will discuss. Instead of solving the four equations (32) he started out with two equations only, assuming a repeat distance d', being one half the actual repeat distance d (Fig. 7). With our notation his solution has the following form:

$$\omega^2(\mathbf{k}) = \omega_0^2 \pm \sqrt{\omega_0^4 - \omega'^4} \tag{41}$$

where

$$\omega_0^2 = \frac{k_s}{m}\left(1 - \cos 2\alpha \cos \frac{d\mathbf{k}}{2}\right) + \frac{2k_b}{m}\left(1 + \cos \frac{d\mathbf{k}}{2}\right)\left(1 + \cos 2\alpha \cos \frac{d\mathbf{k}}{2}\right)$$

$$\omega'^4 = \frac{8k_s k_b}{m^2}\left(1 + \cos \frac{d\mathbf{k}}{2}\right)\sin^2 \frac{d\mathbf{k}}{2}$$

Equation (41) describes only two frequency branches but they extend over twice the regular $1/\lambda$ range (see Figs. 3 and 5) shown in Fig. 10a. The square of the frequency is plotted against $1/\lambda$ in the range from 0 to $1/2d'$.

This abbreviated treatment leads also to four potentially Raman- or infrared-active frequencies for an infinitely long polyethylene chain. They are given by the end points of the two branches (in Fig. 10a) for $1/\lambda = 0$ and $1/2d'$, or $\mathbf{k} = 0$ and π/d'. This can be verified by substituting these two values for $\mathbf{k}$ in Eq. (41) which leads to the four frequencies given by Eqs. (37a–d).

In reality the repeat unit is d and not d' which means that in our more general treatment the $1/\lambda$ range extends only from 0 to $1/2d$. The four frequency branches in this range are shown in Fig. 10b. Now only the frequencies for which $\mathbf{k} = 0$ are potentially Raman- or infrared-active as shown in Chapter II.

Let us compare Fig. 10a with Fig. 10b. The four branches in Fig. 10b can be obtained by "folding" Fig. 10a at a vertical line with $1/\lambda = 1/2d$. This folding

(a)

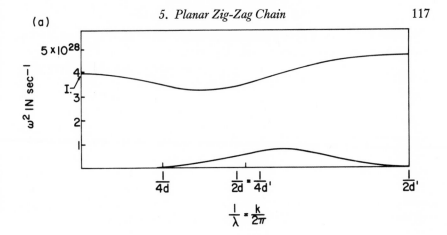

(b)

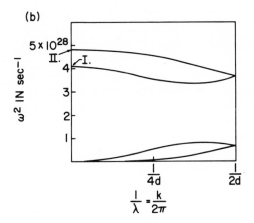

(c)

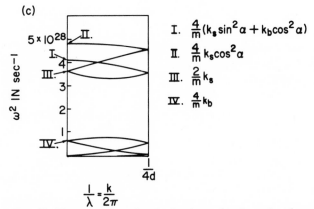

I. $\frac{4}{m}(k_s \sin^2 \alpha + k_b \cos^2 \alpha)$

II. $\frac{4}{m} k_s \cos^2 \alpha$

III. $\frac{2}{m} k_s$

IV. $\frac{4}{m} k_b$

FIG. 10. Frequency branches for in-plane vibrations of a planar zig-zag chain. Repeat unit consists of (a) one point mass, (b) two point masses, (c) four point masses.

operation will be used again in Section 5F to describe chains which have repeat units longer than two carbon atoms.

E. OUT-OF-PLANE VIBRATIONS

Pitzer (Ref. 16) has calculated the out-of-plane vibrational frequencies for a zig-zag hydrocarbon chain. The same method can be applied to a vinyl or vinylidene polymer chain. The out-of-plane frequencies fall below 200 cm^{-1}, as we will see in this section. It has not been possible yet for any polymer to assign definitely an observed absorption in this region to an out-of-plane skeletal vibration. There are two reasons for this: first, this frequency region is not yet easily accessible with conventional infrared instruments, and it is also difficult to obtain Raman spectra for high polymer molecules; second, it turns out that the two potentially active vibrations are zero vibrations, namely, a rotation around the chain axis and a translation perpendicular to the plane of the zig-zag chain, but in an actual polymer other out-of-plane vibrations might be active. They are expected to be rather weak since this activity would be due to deviations from the idealized planar zig-zag chain.

We assume, with Pitzer, that the restoring forces are determined by the torsional force constant k_t for a torsion around a C—C bond. The potential energy of a system of four consecutive point masses is $\frac{1}{2}k_t\Delta z^2$ if any one of the masses has a distance Δz from the plane formed by the other three. The differential equations of motion corresponding to (25) are

$$m_1\ddot{z}_{2n+1}+k_t(z_{2n-2}-2z_{2n-1}-z_{2n}+4z_{2n+1}-z_{2n+2}-2z_{2n+3}+z_{2n+4}) = 0 \quad (42a)$$

$$m_2\ddot{z}_{2n}+k_t(z_{2n-3}-2z_{2n-2}-z_{2n-1}+4z_{2n}-z_{2n+1}-2z_{2n+2}+z_{2n+3}) = 0 \quad (42b)$$

where z_n is the distance of the nth mass from the equilibrium position in the zig-zag plane.

In analogy to (31) we assume a solution of the form

$$z_{2n} = C\exp\left[i\left(\omega t+2n\frac{d\mathbf{k}}{2}\right)\right], \quad z_{2n+1} = C'\exp\left[i\left(\omega t+(2n+1)\frac{d\mathbf{k}}{2}\right)\right] \quad (43)$$

With this solution we can solve (42) and obtain the out-of-plane normal frequencies of a zig-zag chain:

$$\omega^2_{5,6}(\mathbf{k}) = \frac{4k_t\sin^2(d\mathbf{k}/2)}{m_1 m_2}[m_1+m_2\pm\sqrt{m_1^2+m_2^2+2m_1 m_2\cos\overline{d\mathbf{k}}}] \quad (44)$$

The two frequency branches are plotted as dotted lines in Fig. 13a for equal masses and in Fig. 13b for nonequal masses.

Equation (44) reduces to Pitzer's formula (Ref. 16) for a n-hydrocarbon chain where $m_1 = m_2 = m$:

$$\omega^2_{5,6}(\mathbf{k}) = \frac{4k_t}{m}\sin^2\frac{d\mathbf{k}}{2}\left(1\pm\cos\frac{d\mathbf{k}}{2}\right) \quad (45)$$

As we mentioned before one can easily see from (44) and (45) that for an infinitely long chain the potentially active frequencies (where $\mathbf{k} = 0$) are zero vibrations.

If we assume that the selection rules break down and vibrations, where $\mathbf{k} \neq 0$, are active, the highest possible frequencies in each branch would be obtained for $\mathbf{k} = \pi/d$. Their values are

$$\omega_{5\text{max}}^2 = \frac{8k_t}{m_1}, \qquad \omega_{6\text{max}}^2 = \frac{8k_t}{m_2} \qquad (46)$$

Assuming $m_1 < m_2$, all possible frequencies fall into the interval from 0 to $\omega_{5\text{max}}$. For most practical cases this interval extends from 0 to about 200 cm^{-1}.

F. Deviations from a Perfect Zig-Zag Chain

Real vinyl polymer molecules are not infinitely long planar zig-zag chains. Several deviations from this model can occur so that the high symmetry is lost partially and the strict selection rules break down. In the following pages we discuss some of these deviations.

Finite Chain Length

An actual polymer chain might consist of only short straight segments. In an amorphous polymer, for example, an approximately straight segment is only a few monomer units long because of rotation around the C—C bond due to a low energy barrier. Also, for a crystalline polymer straight chain segments are rather short since diameter and length of microcrystals can be as small as 20 to 30 Å and seldom larger than 200 to 300 Å (Refs. 20, 21).

In some cases the finite length of a straight chain segment is determined by the chemical structure. As an example we mention

$$\text{6-nylon--}(\text{-NHCOCH}_2\text{CH}_2\text{CH}_2\text{CH}_2\text{CH}_2\text{-})\text{-}_x$$

where the hydrocarbon chain segment consists of five CH$_2$ groups.

In a finite chain of N repeat units *all* normal vibrations become potentially infrared- or Raman-active. The total number of vibrations $3nN$ (n is the number of atoms per repeat unit) fall on $3n$ frequency branches. For a given frequency branch the absorption intensity decreases very rapidly in going from $1/\lambda = 0$ to $1/\lambda = 1/2d$, so that in many cases only one or two absorption bands of a branch are strong enough to be observed. This problem is discussed in detail in the next chapter.

Repeat Unit Longer Than Two Carbon Atoms

In some vinyl polymers the repeat unit might be longer than in a regular head-to-tail structure (Fig. 11a). For example, it is conceivable to have a head-to-head vinyl polymer (Fig. 11b) with a repeat unit equal to $2d$ containing four carbon atoms. A mathematical treatment (similar to the previous calculations) of such a chain would be very complicated. We therefore assume as an approximation that the normal vibrations are the same as in a two carbon atom repeat unit chain but that the selection rules are different because of the lower symmetry. Since the repeat unit contains twice as many atoms we have twice as many frequency branches but they extend to only one half the $1/\lambda$ range.

The six frequency branches for a chain with unequal masses and a repeat unit d are shown in Fig. 13b. They were calculated numerically as described in Section 5G. For a repeat unit of $2d$ the twelve branches are shown in Fig. 13c. They are obtained by folding Fig. 13b at a vertical line where $1/\lambda = 1/4d$. Now twice as many vibrations are potentially infrared- or Raman-active. In Fig. 13c they are the points of the branches where $1/\lambda = 0$, but in Fig. 13b they would be the points where $1/\lambda = 0$ as well as $1/2d$.

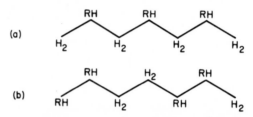

Fig. 11. (a) Head-to-tail structure. (b) Head-to-head structure.

Other Reasons for Lower Symmetry in a Zig-Zag Chain

In vinyl polymers the substituent R (Fig. 11) can be arranged to result in chain segments with either an all d- or an all l-form. Such a polymer is called isotactic, but often the two forms are statistically mixed and the polymer is atactic (Ref. 15).

Two examples for isotactic polymers are polystyrene and polypropylene (Refs. 14, 15). In these two cases the chains are no longer planar. They are coiled into spirals with a regular repeat unit consisting of several monomer units, but the frequencies calculated for an idealized planar vinyl polymer zig-zag chain are probably good approximations for helical polymers (compare Chapter I).

G. Practical Calculations of Skeletal Vibrations

In this section we will make reasonable numerical assumptions for the force constants and the masses for polymers with a zig-zag carbon backbone. First we will calculate the potentially infrared- and Raman-active frequencies for an infinitely long and perfect chain. Then we will show in what frequency range to expect absorption bands for the cases where the selection rules break down.

These calculations have not yet been compared with observed spectra since only tentative assignments are available for a very limited number of polymers. The calculated frequencies should, therefore, serve as guides for future assignments.

Similar calculations of skeletal vibrations can also be carried out on zig-zag backbone chains which do not contain carbon atoms only, such as polyoxymethylene $-(-CH_2-O-)-_n$ which is briefly discussed in the last paragraph of this chapter.

Infinitely Long, Planar Zig-Zag Chains

Only two nonzero vibrations are potentially infrared- or Raman-active. They are the two in-plane vibrations $\omega_2(0)$ and $\omega_4(0)$ shown in Fig. 9. Their values are expressed by Eqs. (37b) and (37d). In the following calculations we shall assume tetrahedral angles for the carbon atom substituents, which means that α (see Fig. 8) becomes 35.25° and

$$\sin^2 \alpha = \tfrac{1}{3}, \qquad \cos^2 \alpha = \tfrac{2}{3} \tag{47}$$

The bending and stretching force constants k_b and k_s are not very accurately known. The results obtained from calculations on small molecules such as ethane, propane, ethanol, etc., vary within a range of about 20% as shown in Table I.

TABLE I

STRETCHING AND BENDING FORCE CONSTANTS FOR A

$\gtrless C - C \lessgtr$ AND A $\gtrless C - \overset{|}{\underset{|}{C}} - C \lessgtr$ GROUP

Compound	$k_s \cdot 10^{-5}$ dynes/cm	$k_b \cdot 10^{-5}$ dynes/cm	Ref.
Ethane	4.30		2
Propane	3.84 (3.78)	0.34 (0.36)	2 (11)
Propane	4.12	0.37	1
Cyclopropane	4.04		2
Isobutane	4.29		2
—	4.50		6
—	4.5–5.6		25

In a vinyl or vinylidene polymer the reduced mass M for a repeat unit ranges from 7 for polyethylene to 14 atomic units for a polymer with an infinitely heavy substituent. For polytetrafluoroethylene the reduced mass is 25.

The expected frequencies $\omega_2(0)$ and $\omega_4(0)$ [Eqs. (29)] were calculated for a range of force constants (listed in Table I) and for various reduced masses. The results are plotted in Fig. 12. The solid lines were obtained with average values for the force constants and the dotted lines give the frequency interval for a 10% uncertainty in the C—C stretching force constant. With the help of this figure it is possible to give the approximate frequencies for the two potentially active skeletal vibrations. For polyvinyl alcohol, for example, the two frequencies are expected to be in the 920 cm^{-1} and in the 1000 cm^{-1} range while polytrifluoroethylene is expected to absorb at about 645 and 695 cm^{-1}. In oriented polymer samples one band should exhibit perpendicular dichroism and the other band parallel dichroism.

Deriving Eqs. (37) we have replaced a CH_2 or a CR_1R_2 group by a point mass. Such an approximation is only permissible if the coupling between the chain carbon atom and the substituents is high compared with the coupling among the chain atoms themselves. Otherwise, the values of m_2, for example (in Fig. 7), will have to be chosen somewhere between the mass of a carbon atom and the mass of a CHR group. This means that the frequency values obtained from Fig. 12 are

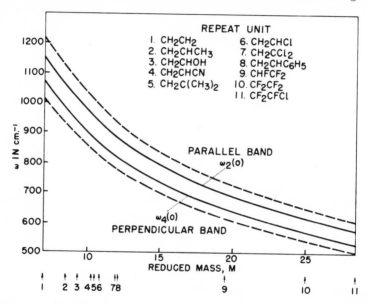

FIG. 12. Potentially infrared- or Raman-active in-plane vibrations of infinitely long zig-zag chains. $k_s = 4.20 \times 10^5$ dynes/cm, $k_b = 0.37 \times 10^5$ dynes/cm. The dashed lines correspond to a 10% variation in the values for the force constants.

lower limits for the actual frequencies since in-chain and side-chain force constants are usually of the same order of magnitude.

The two frequencies are infrared-active for all vinyl and vinylidene polymers, but not in polyethylene or polytetrafluoroethylene. The bands are Raman-active for all polymers as shown in Chapter II.

Breakdown of Strict Selection Rules

In Section 5F various reasons were listed why selection rules can break down and why all vibrations of a frequency branch might become active. It is useful, therefore, to calculate the value of all frequencies in a branch. For this purpose, we have assumed average force constants (see Table I) $k_s = 4.20 \cdot 10^5$ dynes/cm, $k_b = 0.37 \cdot 10^5$ dynes/cm, and $k_t = 0.034 \cdot 10^5$ dynes/cm (Ref. 9) and calculated the

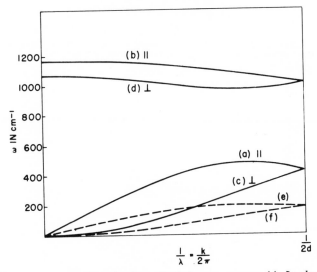

FIG. 13a. Six frequency branches for polyethylene, $m_1 = m_2 = 14$. *In-plane vibrations*: (a) $= \omega_1 (k)$, (b) $= \omega_2 (k)$, (c) $= \omega_3 (k)$, (d) $= \omega_4 (k)$; *out-of-plane vibrations*: (e) $= \omega_5 (k)$, (f) $= \omega_6 (k)$.

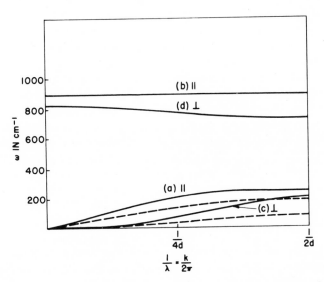

FIG. 13b. Six frequency branches for a planar vinyl polymer zig-zag chain $m_1 = 14$, $m_2 = 83$. KEY: ———: In-plane vibrations; –––: out-of-plane vibrations.

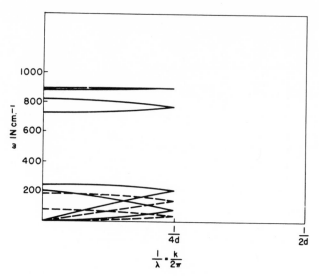

FIG. 13c. Same as Fig. 13b but "folded" at a vertical line where $1/\lambda = 1/4d$. Repeat unit contains 4 chain carbon atoms resulting in 12 frequency branches.

behavior of *all* frequency branches. The in-plane vibrational frequencies were obtained by solving numerically Eqs. (32a–d) with the help of an electronic computer. The two out-of-plane frequency branches are given by Eq. (44). The six branches are plotted in Fig. 13a for polyethylene ($m_1 = m_2 = 14$ atomic units) and in Fig. 13b for polyvinylidene chloride ($m_1 = 14$, $m_2 = 83$). In Fig. 13c are shown the twelve frequency branches for a planar vinyl polymer zig-zag chain, $m_1 = 14$, $m_2 = 83$, with a repeat unit containing four carbon atoms. Six diagrams (Figs. 14a–f) show the results of the numerical computations of the six frequency branches for various zig-zag chains. Most of the curves plotted are for vinyl and vinylidene polymers where $m_1 = 14$ and $m_2 > 14$, but also a few other polymers such as polytetrafluoroethylene (Teflon†), polytrifluorochloroethylene, and polytrifluoroethylene are included.

Figures 14a–d further contain the frequency branches for the skeletal vibrations of polyoxymethylene where we have assumed a planar zig-zag structure with tetrahedral angles. For k_s we used the C—O stretching force constant of dimethyl ether which is $4.53 \cdot 10^5$ dynes/cm (Ref. 11). The bending force constants for a C—O—C and an O—C—O group are about the same (Ref. 11). The value $k_b = 0.35 \cdot 10^5$ dynes/cm was used for the numerical computations. The masses are $m_1 = 14$, $m_2 = 16$. For an infinitely long chain one would expect two frequencies to be infrared-active, resulting in a parallel band at 1171 cm^{-1} and a perpendicular band at 1054 cm^{-1}. Corresponding normal vibrations can also be calculated for helical chain structures (Refs. 12, 13, 22, 23).

† du Pont trademark.

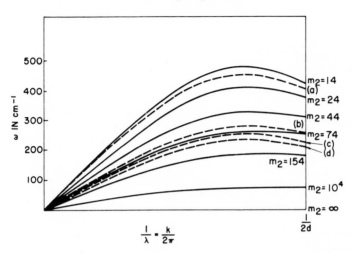

FIG. 14a. Acoustical frequency branches ω_1 (k). KEY: ———: vinyl polymers $m_1 = 14$, m_2 as indicated; ———: (a) $-(-CH_2O-)-_x$, (b) $-(-CHF—CF_2-)-$, (c) $-(-CF_2CF_2-)-_x$, (d) $-(-CF_2CFCl-)-_x$.

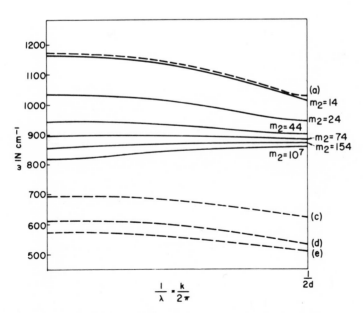

FIG. 14b. Optical frequency branches ω_2 (k). See legend of Fig. 14a.

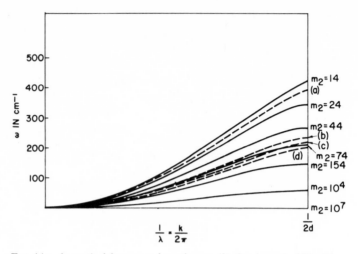

FIG. 14c. Acoustical frequency branches $\omega_3\ (k)$. See legend of Fig. 14a.

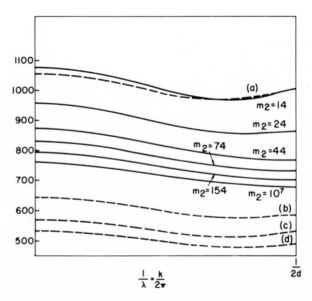

FIG. 14d. Optical frequency branches $\omega_4\ (k)$. See legend of Fig. 14a.

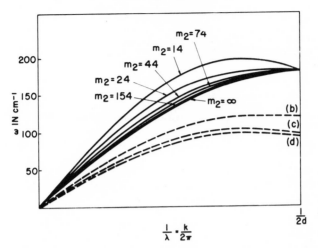

FIG. 14e. Out-of-plane vibrations, acoustical frequency branches, $\omega_5 (k)$. See legend of Fig. 14a.

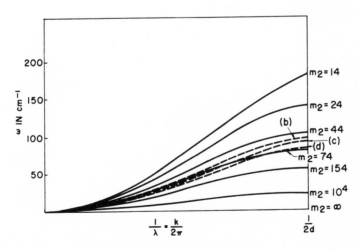

FIG. 14f. Out-of-plane vibrations, acoustical frequency branches, $\omega_6 (k)$. See legend of Fig. 14a.

REFERENCES

1. Ahonen, C. O., *J. Chem. Phys.* **14**, 625 (1946).
2. Bonner, L. G., *J. Chem. Phys.* **5**, 293 (1937).
3. Born, M., and von Kármán, T., *Physik. Z.* **13**, 297 (1912).
4. Brillouin, L., "Wave Propagation in Periodic Structures," 2nd ed. Dover, New York, 1953.
5. De Launay, J., *Solid State Phys.* **2**, 219 (1956).
6. Herzberg, G., "Infrared and Raman Spectra of Polyatomic Molecules." Van Nostrand, Princeton, New Jersey, 1945.
7. Houston, W. V., "Principles of Mathematical Physics." McGraw-Hill, New York, 1948.
8. Kirkwood, J. G., *J. Chem. Phys.* **7**, 506 (1939).
9. Krimm, S., Liang, C. Y., and Sutherland, G. B. B. M., *J. Chem. Phys.* **25**, 549 (1956).
10. Lagrange, J. L., "Méchanique analytique," 4th ed., Vol. 1, p. 405. 1888. (First edition, p. 317. Veuve-Desaint, Paris, 1788.)
11. Mecke, R., and Kerkhof, F., *in* "Landolt-Bornstein, Zahlenwerte und Funktionen, Atom- und Molekularphysik," Vol. I, Part 2, p. 226. Springer, Berlin, 1951.
12. Miyazawa, T., *Spectrochim. Acta* **16**, 1231 (1960).
13. Miyazawa, T., *Spectrochim. Acta* **16**, 1233 (1960).
14. Natta, G., *Makromol. Chem.* **16**, 213 (1955).
15. Natta, G., *J. Polymer Sci.* **16**, 143 (1955).
16. Pitzer, S., *J. Chem. Phys.* **8**, 711 (1940).
17. Rutherford, D. E., *Proc. Roy. Soc. Edinburgh* **A62**, 229 (1947).
18. Rutherford, D. E., *Proc. Roy. Soc. Edinburgh* **A63**, 232 (1951).
19. Shimanouchi, T., and Mizushima, S., *Sci. Papers Inst. Phys. Chem. Res. (Tokyo)* **40**, 467 (1943).
20. Statton, W. O., *J. Polymer Sci.* **28**, 423 (1958).
21. Statton, W. O., and Godard, G. M., *J. Appl. Phys.* **28**, 1111 (1957).
22. Tadokoro, H., *J. Chem. Phys.* **33**, 1558 (1960).
23. Tadokoro, H., Kobayashi, A., Kawaguchi, Y., Sobajima, S., Murahashi, S., and Matsui, Y., *J. Chem. Phys.* **35**, 369 (1961).
24. Wallis, R. F., *Phys. Rev.* **105**, 540 (1957).
25. Wilson, E. B., Decius, J. C., and Cross, P. C., "Molecular Vibrations." McGraw-Hill, New York, 1955.

—IV—

Vibrational Interaction in Chain Molecules

1. The Coupled Oscillator Model

In the previous chapter we have shown how to calculate the vibrational frequencies for certain chains with a well-defined geometry and known force constants. Those calculations were restricted to the skeletal vibrations of a few simple polymer chains, but as soon as group vibrations are included in the treatment the problem gets quite involved. This was shown, for example, by Primas and Günthard (Refs. 29, 30) who calculated all the expected vibrations for normal hydrocarbon chains.

In this chapter we will present an approximation method (Ref. 52) to obtain information about characteristic features in spectra of chain molecules. This method is especially useful if the force constants are not well known or if the geometry is not simple enough to allow a rigorous mathematical treatment.

In this method a characteristic vibration of a single molecular group is represented by a harmonic oscillator. As an example let us consider the C=O stretching vibrations in a 1-nylon chain (Fig. 1a). One monomer unit —CONR— would absorb at about 1660 cm^{-1} which is the characteristic frequency for the amide C=O group. For a long chain of N monomer units we would expect to find a splitting due to the interaction between the C=O vibrations. To calculate the amount of splitting as well as the expected relative absorption intensity of the split components we represent a C=O group by a single oscillator and treat the chain as a one-dimensional set of coupled oscillators for the purpose of studying the spectrum in the C=O region only. For the calculations we will have to assume a certain C=O stretching as well as a coupling force constant. The coupling force constant is small for group vibrations which means that the split bands fall still in the neighborhood of the unperturbed absorption band of a single molecular group.

The same model can be used for any other group frequency, for example, one which is characteristic for the side group R. Even skeletal vibrations can be treated in this fashion but here the coupling force constant is usually of the same order of magnitude as the oscillator force constant itself. The interaction between the oscillators is no longer a perturbation as in the case of characteristic group vibrations.

As a model for a chain molecule let us consider a linear chain of coupled oscillators as shown in Fig. 1b, where each oscillator is an electrical dipole pointing

129

in the direction of the arrows. Without a coupling between the N oscillators we would observe only one infrared absorption band at the frequency ω_0, the frequency of an individual oscillator. But with a vibrational interaction we have a system of coupled oscillators with N degrees of freedom. Such a system has N

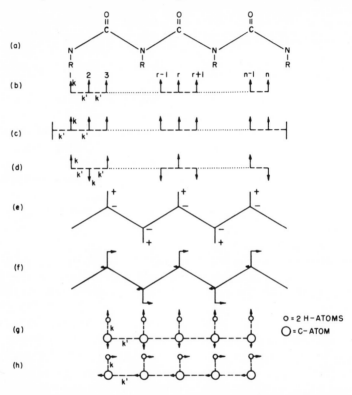

FIG. 1. (a) Planar model of a 1-nylon zig-zag chain. (b) Linear set of N-coupled parallel dipoles with free ends. (c) Linear set of N-coupled parallel dipoles with fixed ends. (d) Linear set of N-coupled antiparallel dipoles with free ends (k = oscillator force constant, k' = coupling force constant). (e) Simplified model of the "all in phase" CH_2-rocking vibration in a n-paraffin chain. (f) Same as (e), but for the CH_2-wagging vibration. (g) Schematic representation (side view) of (e), k is CH_2-rocking force constant (h) Schematic representation (side view) of (f), k is CH_2-wagging force constant.

normal vibrations. We, therefore, would expect a splitting of the absorption band into N components provided all these components are infrared-active. Assuming certain oscillator and coupling force constants one can calculate the amount of this splitting as shown in the following sections. We also will give a method to calculate the expected intensity distribution in the split absorption band.

These calculations will show that for an infinitely long chain only one of the normal vibrations is potentially Raman- or infrared-active. This, of course, is a confirmation of the same result obtained from a group theoretical treatment (see Chapter II). For chains of a finite length many bands in the series may be infrared-active.

What are the advantages and weak points of this method ? A model of coupled oscillators cannot give absolute values for an absorption frequency as opposed to the calculations of Chapter III or the calculations by Primas and Günthard (Refs. 29, 30) on normal paraffins where one starts out with geometrically well-defined models and force constants. On the other hand, the oscillator method is more flexible and can be adapted to a high variety of chain molecules. It can lead to some useful information about the structure of the chain as well as the assignment of absorption bands, especially for cases where a normal coordinate treatment would be too complicated. Group vibrations such as vibrations of CH_2, NO_2, $C=O$, groups, etc., are particularly suited for this treatment. But the assignment of bands may be difficult in a case where two group frequencies of different groups are very close and the split bands overlap.

In the following sections we will carry out the calculations for two specific models. In Section 2 we will calculate the splitting as well as the geometrical form of the normal vibration and in Section 3 the expected infrared activity and band intensities.

2. Normal Vibrations in a Chain of N Coupled Oscillators

A. Introduction

In this section a chain of N coupled oscillators will be treated mathematically. We will assume that each oscillator is an electric dipole† and consider two specific simple models: (a) a set of parallel, and (b) a set of antiparallel dipoles as shown by Figs. 1b and 1d, respectively.

It is important to emphasize that we deal with chains of a *finite* length where the Born–von Kármán boundary condition does not apply. We are interested in the effect of chain length upon the infrared absorption spectrum. As in Chapter III the two cases of free and fixed end chains (see Figs. 1b and 1c) will be discussed side by side.

The mathematical approach described here is similar to the one used in Chapter III. It dates back to Lagrange (Ref. 22) and more recently to Born and von Kármán (Ref. 3). Linear chains of a finite number of oscillators have been analyzed by Routh (Ref. 33) and more specifically by Parodi (Ref. 27), Rutherford (Ref. 34), and by Whitcomb, Nielsen and Thomas (Ref. 51). More elaborate treatments involving interactions between nearest, second-nearest, etc., neighbors were given by Deeds (Ref. 15) and Brillouin (Ref. 5). Brown, Sheppard, and Simpson (Refs. 7, 8) and Tschamler (Ref. 48) have used this approach in the interpretation of the infrared and Raman spectra of normal paraffins.

---

† It is not necessary to assume a change of dipole moment with the motion of an oscillator in order to calculate the normal vibrations of a chain. But it is convenient to do so since without any dipole moment change a vibration could not be infrared-active.

B. Chain of Parallel Dipoles

The C=O stretching vibration in 1-nylon (Fig. 1a) and the CH₂-rocking and -wagging vibrations in normal paraffins (Figs. 1e and 1f) are examples of chains where for a specific motion in a group vibration all dipole moment changes are parallel. Figures 1e and 1f are simplified models and Figs. 1g and 1h are schematic representations for chains of parallel harmonic oscillators with a reduced mass m, a force-constant k, and a coupling force constant k'.

We now will set up the equations of motion for chains with free and fixed ends as shown in Figs. 1b and 1c, respectively. Considering nearest neighbor interaction only, the N equations of motion for the N oscillators are:

Free ends

$$m\ddot{x}_1 = k(-x_1) + k'(x_2 - x_1)$$

$$\vdots$$

$$m\ddot{x}_r = k'(x_{r-1} - x_r) + k(-x_r)$$
$$+ k'(x_{r+1} - x_r)$$

$$\vdots$$

$$m\ddot{x}_N = k'(x_{N-1} - x_N) + k(-x_N)$$

Fixed ends

$$m\ddot{x}_1 = k'(-x_1) + k(-x_1)$$
$$+ k'(x_2 - x_1)$$

$$\vdots$$

$$m\ddot{x}_r = k'(x_{r-1} - x_r) + k(-x_r)$$
$$+ k'(x_{r+1} - x_r) \qquad (1a, b)$$

$$\vdots$$

$$m\ddot{x}_N = k'(x_{N-1} - x_N) + k(-x_N)$$

where x_r is the deviation of the rth oscillator from its equilibrium position.

The signs in Eqs. (1a, b) are chosen in an arbitrary but consistent way so that the effect of a change in amplitude of oscillator $r-1$ and $r+1$ on oscillator r is the same. If one oscillator changes its amplitude in the positive direction, the coupling force causes the adjacent one to move alike (this situation would be reversed for the case where adjacent oscillators point in opposite directions as we will see in the next subsection). The coupling force is assumed to be proportional to the difference in amplitude of two adjacent oscillators.

Equations (1a, b) are mathematically equivalent with Eqs. (1a, b) of the previous chapter. We assume a periodic solution

$$x_r = x_{r,0}\, e^{i\omega t} \qquad (2)$$

and introduce

$$\omega_0^2 = \frac{k}{m}, \qquad \omega'^2 = \frac{k'}{m} \qquad (3)$$

where ω_0 is the frequency of the uncoupled oscillator and ω' is an interaction parameter which has the dimension of a frequency. With (2) and (3), Eqs. (1a, b) become

Free ends

$$(B+1)x_1 + x_2 \quad = 0$$

$$\vdots$$

$$x_{r-1} + Bx_r + x_{r+1} = 0$$

$$\vdots$$

$$x_{N-1} + (B+1)x_N = 0$$

Fixed ends

$$Bx_1 + x_2 \qquad\quad = 0$$

$$\vdots$$

$$x_{r-1} + Bx_r + x_{r+1} = 0 \qquad (4a, b)$$

$$\vdots$$

$$x_{N-1} + Bx_N \qquad\quad = 0$$

B is a dimensionless parameter related to the vibrational frequency as follows:

$$B = \frac{\omega^2 - \omega_0^2}{\omega'^2} - 2 \tag{5}$$

or

$$\omega^2 = (B+2)\omega'^2 + \omega_0^2 \tag{6}$$

The systems of homogeneous equations (4a, b) have solutions only if their secular determinants disappear, that is, if

$$
\begin{vmatrix}
B+1 & 1 & 0 & . & . & . & 0 \\
1 & B & 1 & 0 & . & . & 0 \\
0 & 1 & B & 1 & 0 & . & 0 \\
\vdots & & & & & & \\
0 & & . & . & . & 0 & 1 & B+1
\end{vmatrix} = 0
\quad
\begin{vmatrix}
B & 1 & 0 & . & . & . & 0 \\
1 & B & 1 & 0 & . & . & 0 \\
0 & 1 & B & 1 & 0 & . & 0 \\
\vdots & & & & & & \\
0 & & . & . & . & 0 & 1 & B
\end{vmatrix} = 0 \tag{7a, b}
$$

These determinants can be evaluated (see Ref. 3), and Eqs. (7a, b) assume the form

$$\frac{2 \sin N\theta(1+\cos\theta)}{\sin\theta} = 0 \quad \left| \quad \frac{\sin(N+1)\theta}{\sin\theta} = 0 \right. \tag{8a, b}$$

where θ is defined by

$$2\cos\theta = B \tag{9}$$

The N solutions of (8a, b) are

$$\theta_s = \frac{s\pi}{N} \quad \left| \quad \theta_s = \frac{s\pi}{N+1} \right. \tag{10a, b}$$

where $s = 1, 2, \ldots, N$. From Eqs. (6), (9), and (10) we obtain the N discrete values for the vibrational frequencies ω_s:

$$\omega_s^2 = \omega_0^2 + 2\omega'^2\left(1+\cos\frac{s\pi}{N}\right) \quad \left| \quad \omega_s^2 = \omega_0^2 + 2\omega'^2\left(1+\cos\frac{s\pi}{N+1}\right) \right. \tag{11a, b}$$

$$= \omega_0^2 + 4\omega'^2\sin^2\frac{(N-s)\pi}{2N} \quad \left| \quad = \omega_0^2 + 4\omega'^2\sin^2\frac{(N+1-s)\pi}{2(N+1)} \right. \tag{12a, b}$$

The distribution of frequencies according to Eqs. (11a, b) is shown in Figs. 2a and 2b, respectively (see also Refs. 7, 8, 48), where for convenience the square of the frequency is plotted. All the ω_s^2 values lie in an interval which has a width of $4\omega'^2$. The edges of the interval are given by the values ω_0^2 and $\omega_0^2 + 4\omega'^2$. For an infinitely long chain (11a) and (11b) are equivalent.

If we introduce the wave-number vector $\mathbf{k}_s$ as defined by Eqs. (8) and (9) in

(a)

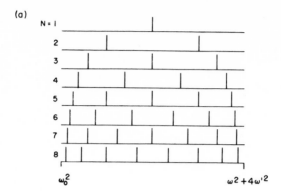

(b)

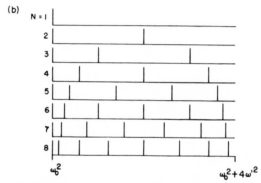

FIG. 2. Normal vibrational frequencies for a chain of N-coupled oscillators $N = 1, 2, \ldots, 8$. (a) Fixed ends; (b) free ends.

the previous chapter, we can express (12a, b) in the following form which is correct for free and fixed ends:

$$\omega_s^2 = \omega_0^2 + 4\omega'^2 \sin^2 \frac{d\mathbf{k}_s}{2} \tag{13}$$

where d is the distance between two oscillators. $\mathbf{k}_s$ can be replaced by a continuous wave-number vector $\mathbf{k}$ which makes (13) an equation with continuous variables ω and $\mathbf{k}$:

$$\omega^2 = \omega_0^2 + 4\omega'^2 \sin^2 \frac{d\mathbf{k}}{2} \tag{13a}$$

ω^2 is plotted as a function of $\mathbf{k}$ in Fig. 3. This frequency branch is a representative curve for free as well as fixed end chains of any length N, but the distribution of the actual frequencies on the branch is different for different cases. This distribution is shown in Fig. 3 for two examples. The frequencies for a free end chain are plotted as squares and for a fixed end chain as circles.

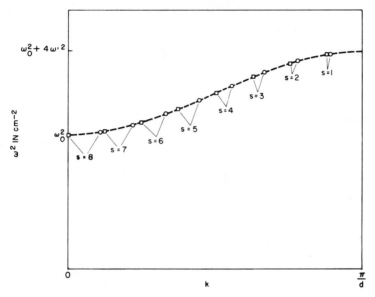

FIG. 3. Frequency branch for a chain of parallel coupled dipoles. The points correspond to $N = 8$. KEY: □: free ends; ○: fixed ends.

C. CHAIN OF ANTIPARALLEL DIPOLES

Figure 1d shows an example for such a chain. It represents schematically CH_2-bending or -symmetric stretching vibrations of a normal hydrocarbon chain.

The problem is very similar to the one we solved previously. We again have a set of coupled oscillators but adjacent ones point in opposite directions.

As an illustration we consider the case of fixed end groups. The system of equations of motion is the same as Eq. (1) except for the signs. For the rth dipole we have

$$\vdots$$
$$m\ddot{x}_r = -k'(x_{r-1} + x_r) - kx_r - k'(x_{r+1} + x_r) \qquad (14)$$
$$\vdots$$

where x_r is again the deviation of the rth oscillator from its equilibrium position. It (x_r) is positive if the hydrogen atoms move away from the chain.

Equation (14) can be written in a form equivalent to (5) if B is replaced by $-B$. Then the expressions for the N frequencies ω_s for an alternating chain become

Free ends	Fixed ends
$\omega_s^2 = \omega_0^2 + 2\omega'^2\left(1 - \cos\dfrac{s\pi}{N}\right)$	$\omega_s^2 = \omega_0^2 + 2\omega'^2\left(1 - \cos\dfrac{s\pi}{N+1}\right)$ (15a, b)

It is obvious from Eqs. (11) and (15) that in both cases the N frequencies as a set are identical although the two frequencies corresponding to the same integer number s are different.

D. Magnitude of Expected Splitting in the Spectrum

According to Eqs. (11) and (15) the values of the split frequencies fall into the interval with the limits ω_0 and $(\omega_0^2 + 4\omega'^2)^{1/2}$. If the interaction force constant is small, that is, if

$$\omega' \ll \omega_0 \qquad (16)$$

the width of this interval $\Delta\omega_0$ can be expressed in the following form:

$$\Delta\omega_0 \approx \frac{2\omega'^2}{\omega_0} \qquad (17)$$

If we introduce the wavelength λ by

$$\lambda = \frac{1}{\omega} \qquad (18)$$

where λ is expressed in cm and ω in cm^{-1}, the wavelength interval $\Delta\lambda_0$ covered by the split band is

$$\Delta\lambda_0 = \frac{\Delta\omega_0}{\omega_0^2} = \frac{2\omega'^2}{\omega_0^3} \qquad (19)$$

For a practical example where $\omega' = 100$ cm^{-1}, the splitting has been calculated for two values of ω_0. The results are listed in Table I. λ_0 and $\Delta\lambda_0$ are expressed in μ.

TABLE I

ω_0 (cm^{-1})	λ_0 (μ)	ω' (cm^{-1})	$\Delta\omega_0$ (cm^{-1})	$\Delta\lambda_0$ (μ)
600	16.667	100	33.3	0.925
3000	3.333	100	6.7	0.0074

The splitting expressed in microns increases with the third power of the wavelength for a given interaction force constant ω'. It is, therefore, much more likely to observe splittings for long-wavelength bands than for short-wavelength bands. This is in agreement with observed spectra as shown in Section 4.

E. Geometrical Form of Normal Vibrations

The geometrical form of a vibration ω_s is given by the displacement x_r^s of each oscillator r from its equilibrium position. It can be determined in the following way.

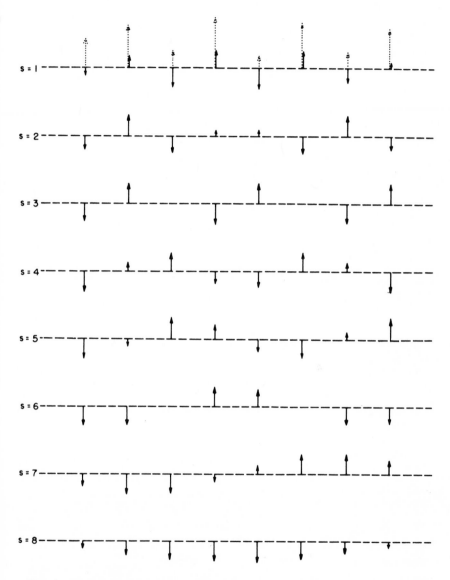

FIG. 4a. Linear set of 8 coupled parallel dipoles with fixed ends. KEY: ———: change of dipole moment for each individual oscillator. ......: actual dipole moment for each individual oscillator only shown for $s = 1$.

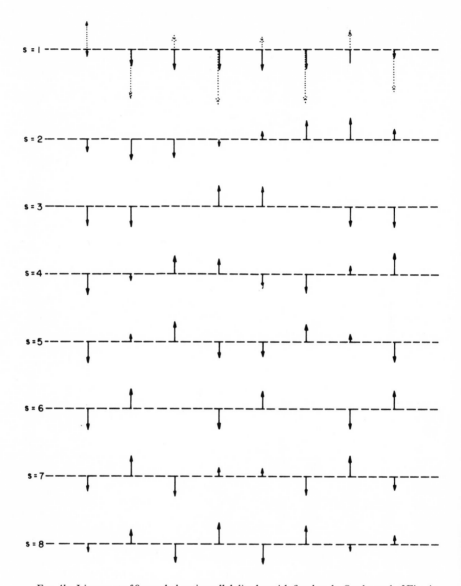

FIG. 4b. Linear set of 8 coupled antiparallel dipoles with fixed ends. See legend of Fig. 4a.

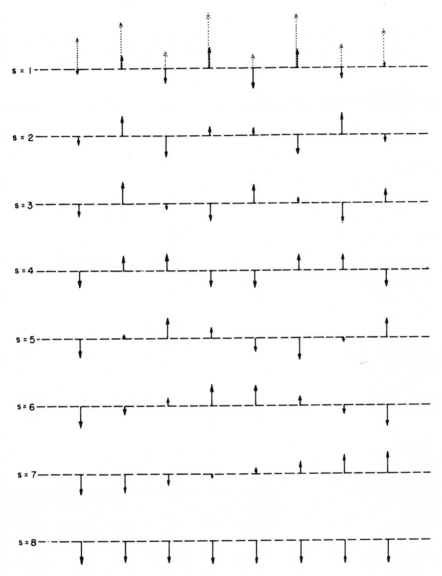

FIG. 4c. Linear set of 8 coupled, parallel dipoles with free ends. See legend of Fig. 4a.

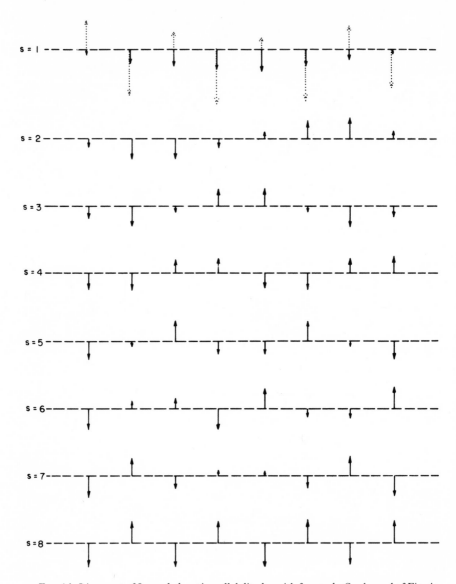

FIG. 4d. Linear set of 8 coupled, antiparallel dipoles with free ends. See legend of Fig. 4a.

Equation (4) is a system of N linear homogeneous equations for the N variables x_r $(r = 1, 2, \ldots, N)$. It has a solution only if its secular determinant disappears, that is, if B assumes the values B_s $(s = 1, 2, \ldots, N)$, given by Eqs. (9) and (10). Under these conditions (4) becomes

Free ends	Fixed ends

$$\left(2\cos\frac{s\pi}{N}+1\right)x_1+x_2 \quad = 0 \qquad \Big| \qquad 2\cos\frac{s\pi}{N+1}x_1+x_2 \qquad = 0$$

$$\vdots \qquad\qquad\qquad\qquad\qquad\qquad \vdots$$

$$x_{r-1}+2\cos\frac{s\pi}{N}x_r+x_{r+1} = 0 \qquad x_{r-1}+2\cos\frac{s\pi}{N+1}x_r+x_{r+1} = 0 \quad (20a, b)$$

$$\vdots \qquad\qquad\qquad\qquad\qquad\qquad \vdots$$

$$x_{N-1}+\left(2\cos\frac{s\pi}{N}+1\right)x_N = 0 \qquad x_{N-1}+2\cos\frac{s\pi}{N+1}x_N \qquad = 0$$

This system of homogeneous equations is mathematically equivalent to Eq. (III,6) and its solution has, therefore, the form of (III,7):

$$x_r^s = A_s(-1)^r\sin\frac{s(2r-1)}{2N}\pi \qquad \Big| \qquad x_r^s = A_s(-1)^r\sin\frac{sr}{N+1}\pi \qquad (21a, b)$$

x_s^r is the deviation from the equilibrium position of the rth dipole for a normal vibration ω_s, and A_s is the normal coordinate.

Equations (20) and (21) hold for a chain of parallel as well as antiparallel dipoles. The sign of x_r^s is chosen in a consistent way. If during a vibration the dipole moment increases x_r^s is positive.

The form of the vibrations according to Eq. (21) is shown in Fig. 4 for the case of eight dipoles. A comparison of this figure with Fig. 2 of the previous chapter shows the close relationship between a chain of point masses and a chain of coupled dipoles.

3. Infrared Spectra of a Finite Chain of Coupled Oscillators, Selection Rules, and Expected Intensity Distribution in Progression Band Series

A. General Discussion

A chain of N coupled dipoles has N normal vibrations. Their frequencies are given by Eqs. (11a, b) for a chain of parallel dipoles and by (15a, b) for a chain where adjacent dipoles are antiparallel.

In this section we will discuss the infrared absorption spectra of such dipole chains. We will show which of the N vibrations are infrared-active, and calculate in a classical way the relative absorption intensity among the active bands.

To explain the method let us consider the classical example of two coupled dipoles pointing in the same direction. This system has two normal vibrations.

For one of them the dipoles vibrate in phase. In the other case they are out of phase by 180°. The in-phase mode is infrared-active since the total dipole moment of the system changes during the vibration. In the out-of-phase mode the dipole moment changes of the two dipoles cancel each other and the vibration is, therefore, not infrared-active.

The infrared activity for the vibrations in a chain of dipoles can be discussed in an analogous way. The absorption intensity is proportional to the square of the change of the total dipole moment during a vibration.† We therefore will first calculate the relative change of the total dipole moment for the N vibrations in a chain of N coupled dipoles. This will give us the expected absorption intensity distribution among the N frequencies.

B. Total Dipole Moment Change for a Normal Vibration

The dipole moment change for the rth dipole during a vibration ω_s is proportional to the change in amplitude x_r^s given by Eqs. (21a, b). For a given vibration ω_s, the relative total dipole moment change ΔM_s can be obtained by adding up the x_r^s values as vectors. ΔM_s can, therefore, be expressed in the following way:

$$\Delta M_s = \sum_{r=1}^{N} x_r^s \qquad \text{all dipoles pointing in the same direction} \qquad (22)$$

$$\Delta M_s = \sum_{r=1}^{N} (-1)^r x_r^s \qquad \text{adjacent dipoles pointing in opposite directions} \qquad (23)$$

The results of these summations are given below‡ for parallel and antiparallel dipoles.

1. Parallel dipoles

From Eqs. (21a, b) and (22) we obtain

$$\Delta M_s = \sum_{r=1}^{N} (-1)^r \sin \frac{s(2r-1)}{2N} \pi \qquad \Delta M_s = \sum_{r=1}^{N} (-1)^r \sin \frac{sr}{N+1} \pi$$

$$= 0, \qquad \text{if } s \neq N \qquad\qquad = 0, \qquad \text{if } s+N \text{ is odd} \qquad (24a, b)$$

$$= -N, \qquad \text{if } s = N \qquad\qquad = -\tan \frac{s\pi}{2(N+1)},$$
$$\qquad\qquad\qquad\qquad\qquad\qquad\qquad \text{if } s+N \text{ is even}$$

---

† This is only true as long as we have a coherent absorption, that is, as long as the wavelength of the radiation is large compared with the dimensions of the absorbing system.

‡ The constant factor A_s from Eq. (21) is omitted since we are interested only in relative values.

2. Antiparallel dipoles

Equations (21a, b) and (23) can be combined to give

$$\Delta M_s = \sum_{r=1}^{N} \sin\frac{s(2r-1)}{2N}\pi \qquad\qquad \Delta M_s = \sum_{r=1}^{N} \sin\frac{sr}{N+1}\pi$$

$$= 0, \qquad\text{if } s \text{ is even} \qquad\qquad = 0, \qquad\text{if } s \text{ is even} \qquad (25a, b)$$

$$= \left(\sin\frac{s\pi}{2N}\right)^{-1}, \quad\text{if } s \text{ is odd} \qquad\qquad = \cot\frac{s\pi}{2(N+1)}, \\ \text{if } s \text{ is odd}$$

$$s = 1, 2, \ldots, N$$

For parallel dipoles with free ends, only the vibration where all dipoles move in phase is infrared-active. In the other three cases $N/2$ or $(N+1)/2$ vibrations are infrared-active, depending on whether N is even or odd.

C. Absorption Intensity Distribution in Band Series

The absorption intensity of a vibration ω_s is proportional to the square of the total dipole moment change ΔM_s.

The absorption process can be treated as a "forced vibration" of the system. By this more refined approach it can be shown that the absorption is not only proportional to ΔM_s^2 but also to the square of the frequency ω_s. Therefore, the relative band intensity I_s becomes

$$I_s = \omega_s^2 \Delta M_s^2 \qquad (26)$$

In most practical cases the frequency range as shown in Fig. 2 is rather narrow, and the factor ω_s^2 in Eq. (26) is practically a constant over the width of an absorption band series. This is particularly true if the coupling force constant k' is small compared with the oscillator force constant k.

From Eqs. (24)–(26) we obtain the equations for the expected absorption intensity distribution in a band series for parallel dipoles:

Free ends	Fixed ends
$I_s = 0,$ if $s \neq N$	$I_s = 0,$ if $s + N$ is odd
$I_s = \omega_s^2 N^2,$ if $s = N$	$I_s = \omega_s^2 \tan^2 \dfrac{s\pi}{2(N+1)},$ if $s + N$ is even

(27a, b)

for antiparallel dipoles:

$I_s = 0,$ if s is even	$I_s = 0,$ if s is even
$I_s' = \omega_s^2 \left(\sin^2 \dfrac{s\pi}{2N}\right)^{-1},$ if s is odd	$I_s = \omega_s^2 \cot^2 \dfrac{s\pi}{2(N+1)},$ if s is odd

(28a, b)

These general equations correspond to those derived by Vidro and Volkenstein (Ref. 49) for skeletal vibrations of normal hydrocarbon chains (Ref. 19).

The intensity distribution in the band series according to Eqs. (27) and (28) is shown in Figs. 5a–d for eight dipoles. In every case (for parallel as well as anti-parallel dipoles) the lowest frequency band is very strong compared to the other

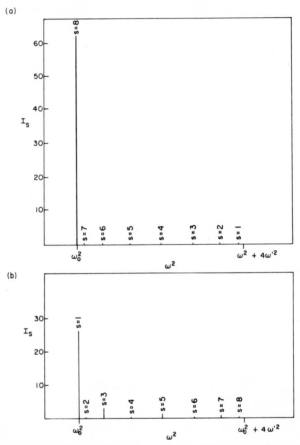

FIG. 5a, b. Expected infrared absorption spectrum of 8 coupled dipoles with free ends. (a) Parallel dipoles; (b) antiparallel dipoles.

bands in the series. The intensity of the second strongest band is never higher than 25%, and it can be as low as 3% of the intensity of the strongest band. Only the strongest band is infrared-active in the case of parallel dipoles with free ends.

For sharp bands and a high splitting it should be possible to observe every single infrared-active absorption band, although the bands at the high frequency end of the branch might be too weak to be detected. This is especially so for fixed end chains where the intensity drops off very rapidly toward the high frequency side of the branch. If the single bands cannot be resolved one should observe a broad absorption band having the shape of the envelope of the band series as shown by the dotted line in Fig. 5d. The expected absorption spectrum for such a case is

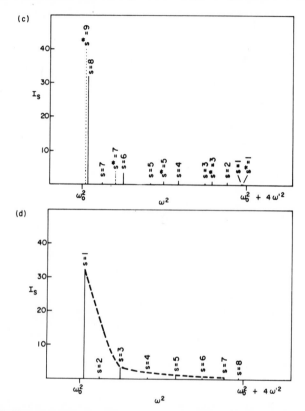

FIG. 5c, d. Expected infrared absorption spectrum of 8 coupled dipoles with fixed ends. (c) Parallel dipoles, the 5 active bands for $N = 9$ are also shown as dotted lines and marked with an asterisk; (d) antiparallel dipoles.

given schematically in Fig. 6. It shows the spectral distribution obtained by scanning infinitely sharp bands with a triangular slit function. Two curves are plotted for a set of eight antiparallel dipoles with fixed and free ends, respectively. The peak intensities are adjusted to the same height. The bands are asymmetric. They are shaded towards higher frequencies.† The wing intensity for a free end chain is higher than for a chain with fixed ends as one would expect by comparing Eqs. (28a) and (28b).

† A symmetric band is expected, of course, for a set of parallel dipoles with free ends where only the band for which $s = N$ is infrared-active.

The effect of increasing chain length is shown in Fig. 7 where the expected band series for a chain of 35 antiparallel dipoles with fixed ends is given. Only the infrared-active bands are plotted. The strongest band ($s = 1$) shifts toward the frequency ω_0 and reaches the value asymptotically for an infinitely long chain. For a chain with free ends, this band appears always at the frequency ω_0 independent of chain length.

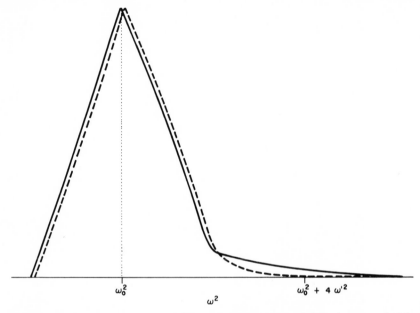

FIG. 6. Expected absorption band series of 8 coupled antiparallel dipoles scanned with a triangular slit-function. KEY: ———: Free ends; – – –: fixed ends.

Increasing the chain length makes it more difficult to resolve the increasing number of bands in the frequency interval ω_0 to $(\omega_0^2 + 4\omega'^2)^{1/2}$ and a single, asymmetric broad band is observed as described above. This broad band becomes sharper if the chain gets longer which is obvious from a comparison of Fig. 5d with Fig. 7. For an infinitely long chain one would expect to observe only one sharp and symmetric band at ω_0. This result has already been obtained in Chapter II where normal vibrations of infinitely long chains were considered from a group theoretical point of view.

4. Comparison of Theory with Observed Spectra

A. LIMITATIONS OF THE THEORETICAL TREATMENT

Any theoretical treatment of a vibrating chain can only lead to semiquantitative and sometimes only to qualitative results, since one has to assume a certain model for a chain or a chain segment, and any model is only an approximation to reality. Some of the major deviations from an idealized model are discussed in the following paragraphs. They will have to be kept in mind if one wants to apply theoretical results to the interpretation of observed spectra.

For the coupled oscillator method we have assumed a linear arrangement of equivalent groups along a chain, but only a limited number of chain molecules in the crystalline state are linear in reality.

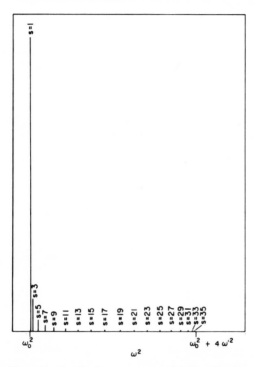

FIG. 7. Expected intensity distribution among the 18 infrared-active components of the progression band series for a linear set of 35 antiparallel, coupled dipoles with fixed ends.

In our calculations we did not deal with vibrational interactions between adjacent chains. Such interactions can cause considerable band shifts and splittings which are usually different in amorphous and crystalline samples.

In the previous sections we treated vibrations of chains with fixed and free ends. The fixed end treatment corresponds to chains with infinitely heavy end groups while a free end chain does not have any end groups at all. A real chain will correspond to an intermediate model.

If the end groups are highly polar they can affect the expected intensity distribution in an absorption band series as we will see in the case of the CH_2-twisting and -wagging vibrations of long chain carboxylic acids.

We now will discuss some examples of spectra which can be partially interpreted by applying the theories developed in the previous sections. Our particular interest will be in band shifts and splittings into progression band series as first observed by Jones, McKay, and Sinclair (Ref. 21) and by Brown, Sheppard, and Simpson (Refs. 7, 8).

B. Hydrocarbon Zig-Zag Chains

Let us consider hydrocarbon chains of the form R_1—$(CH_2)_N$—R_2 where R_1 and R_2 can be any kind of end groups. We will particularly be interested in the absorption spectrum caused by the CH_2 groups. The vibrational modes of a single CH_2

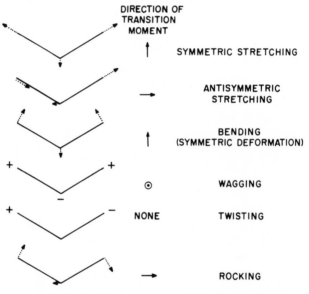

FIG. 8. Vibrational modes of a $\rangle CH_2$ group.

group are shown in Fig. 8. The direction of the dipole moment change that occurs during the vibrations is indicated in the same figure. All vibrational modes except the twisting vibration should be infrared-active.

In a finite chain we replace every individual CH_2 group by an oscillator with the vibrational frequency of one of the modes shown in Fig. 8. The expected spectrum of such a set of coupled oscillators is spread over a frequency range as discussed in the previous sections. For another vibratioanl mode a CH_2 group is again replaced by an oscillator of the corresponding frequency and treated separately as a different set of coupled oscillators. Interaction between two such sets is neglected. In the following subsections we will show that such a treatment can give at least a semiquantitative understanding of the observed spectra of hydrocarbon chains. These spectra have been investigated by many workers (see, e.g., Refs. 6, 7, 28, 29, 30, 36, 39, 48) [an excellent summary of this subject was given by Sheppard (Ref. 35)], but there is still some ambiguity about the assignments of all observed absorption bands. Figure 9 gives the results of such assignments by a number of authors. The horizontal lines cover the frequency range for various CH_2 vibrations. Included are also the carbon-carbon skeletal vibrations. The CH_2-stretching vibrations are not shown since there is no doubt about their assignment to the bands in the 2800 cm^{-1} region.

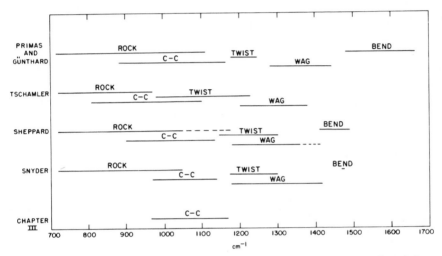

FIG. 9. Frequency ranges for CH_2 and skeletal vibrations of normal hydrocarbon chains.

If a compound is amorphous (solid, liquid, solution, or vapor), the CH_2 groups do not form very long, straight, planar zig-zag chains because the energy barrier for rotation around a C—C bond is very low. In such a case the theory of a linear set of coupled dipoles can only be applied as an approximation to interpret the absorption spectrum. We, threefore, will mainly consider straight segments of amorphous and crystalline compounds.

In the crystalline state there is a vibrational interaction between adjacent chains which usually influences the spectrum. Therefore, we will try to separate interactions within the chain from interactions between chains. The effect of crystallinity on the spectrum has already been discussed in Chapters I and II.

CH_2-*Rocking Vibration*

Hydrocarbon chains show a strong absorption in the 725 cm^{-1} region which has been assigned to the CH_2-rocking vibration by Sutherland and Sheppard (Ref. 43). The dipole moment of a single CH_2 group changes perpendicular to the plane of

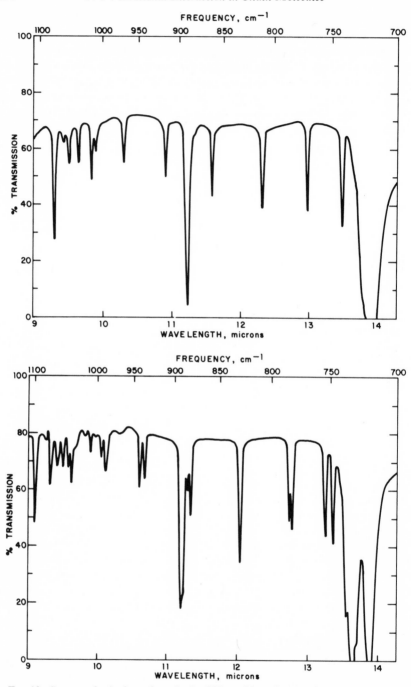

FIG. 10. Spectra of *n*-hydrocarbons in the CH₂-rocking vibrational region, at −160° C, after Snyder (Ref. 38). (a) C₂₄H₅₀; (b) C₂₃H₄₈.

the carbon zig-zag chain (Fig. 8). The vibrational mode where all CH_2 groups rock in the same direction simultaneously is shown in Fig. 1e. This means that the set of parallel oscillators shown in Figs. 1b and 1g is the adequate model. As an example, an absorption spectrum in the CH_2-rocking vibrational region is shown in Fig. 10a for the normal hyrdocarbon $C_{24}H_{50}$ in the crystalline state (triclinic form) at $-160°$ (Ref. 38). This spectrum shows a number of the components of the progression band series between 700 and 1000 cm^{-1}. First we will discuss the frequency and then the intensity of the individual components of the band series.

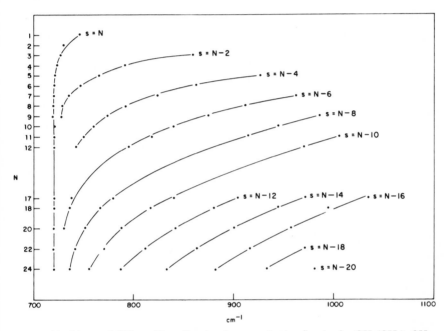

FIG. 11. Observed CH_2-rocking vibrational progression band series for $CH_3(CH_2)_N CH_3$ (Refs. 8, 39).

Figure 11 gives the frequencies of the observed rocking vibrational absorptions for a large number of normal hydrocarbons. The data in this figure were obtained by Brown, Sheppard, and Simpson (Ref. 8) and by Snyder (Ref. 39). We now compare the observed frequency distribution in the progression band series with theoretical predictions. For the coupled oscillator model the frequencies are given by Eq. (11a) for a free end chain and by Eq. (11b) for a chain with fixed ends. Comparing Fig. 11 with Figs. 2a and 2b we see that the fixed end model is some-what more realistic since for the free end chain the lowest frequency ($s = N$) would have to be the same for all progression band series independent of chain length. The frequency of all absorption bands can be expressed as a simple function of the number of oscillators N and the parameter s. In Fig. 12 the square of the observed frequency ω_s is plotted as a function of $1 + \cos\left[s\pi/(N+1)\right]$ for six chains with $N = 6, 9, 12, 17, 22,$ and 24. We indeed observe that all the points fall on a smooth curve, indicating that their position can be described by one and the same equation; but this curve is not a straight line as we would expect from Eq. (11b). This equation can, therefore, be considered only a first approximation for the actual frequency distribution. The lower frequency of the CH_2-rocking vibrational range is 720 cm^{-1} and corresponds to ω_0 while the upper frequency is about 1047 cm^{-1} and corresponds to $(\omega_0^2 + 4\omega'^2)^{1/2}$ [see Eq. (11)]. From this we

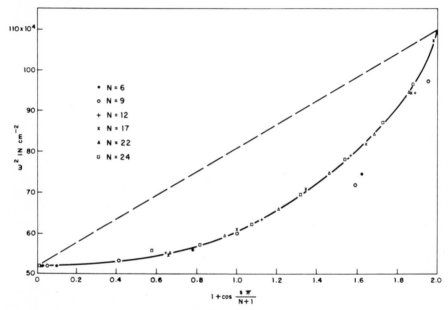

FIG. 12. Square of the observed CH_2-rocking vibrational bands for normal hydrocarbons plotted against $1 + \cos\left[s\pi/(N+1)\right]$, corresponding to a fixed end model with N coupled oscillators.

calculate an average value for the interaction parameter ω' of 381 cm^{-1}. The main reason for the deviation of the curve in Fig. 12 from a straight line is the fact that there is a considerable crowding of absorption bands towards the low frequency end of the progression band series. This fact was predicted by Deeds (Ref. 15) from a more rigorous treatment of coupled oscillators including nearest neighbor as well as next to nearest neighbor interactions. The results of a normal coordinate analysis (Ref. 29) are discussed at the end of this subsection.

The intensity distribution in the band series agrees qualitatively with the expected behavior. It is particularly apparent in Fig. 10a that the lowest frequency band is by far the strongest one in the series. The intensity of the remaining bands falls off towards higher frequencies although not quite as rapidly as one would expect from a tangent square law [Eq. (27b) and Fig. 5c]. In a free end chain of parallel dipoles only the strongest band in the series would be infrared-active. But this is not the case in the observed spectrum, which is another indication that the fixed end model comes closer to the real molecule than the free end model.

For the n-$C_{23}H_{48}$ chain the general behavior of the band series is very similar to the case of n-$C_{24}H_{50}$, but most of the bands in the series are split due to the interaction between adjacent chains (see Fig. 10b). The fact that the splitting occurs only in some hydrocarbons and not in others is due to the difference in crystal structure. $C_{23}H_{48}$ is orthorhombic (large splitting) while $C_{24}H_{50}$ has the

TABLE II

CH₂ ROCKING ABSORPTION FREQUENCIES IN cm^{-1} (REFS. 24, 53)

Hydrocarbon	N	Gas	Liquid
Propane	1	748	
Butane	2	732	
		741 ?	
		748 ?	
Pentane	3	730	732
			726
Hexane	4		725
Heptane	5		723
Octane	6	723	722
Nonane	7		721
Decane	8		721
Tetradecane	12		721
2,2-Dimethylpentane	2		741
2,2,5,5-Tetramethylhexane	2		755
2,2-Dimethylhexane	3		728
2,2-Dimethylheptane	4		725

triclinic crystal form and does not show the splitting (Refs. 1, 9, 23, 37, 38). This splitting was also discussed in connection with polyethylene (Chapter II).

If the spectra of $C_{24}H_{50}$ and $C_{23}H_{48}$ are measured with the sample at room temperature, only the strongest absorption band in the rocking vibrational band series can be observed. This temperature effect has been discussed in detail in Chapter II.

According to the coupled oscillator model the position of the strongest band should shift as a function of chain length in the following way. From Eq. (11b) or (15b) we obtain for fixed ends

$$\omega_{\text{strongest}}^2 = \omega_0^2 + 2\omega'^2\left(1 + \cos\frac{N\pi}{N+1}\right) \tag{29}$$

Table II lists the strongest bands in the 720–750 cm^{-1} region for normal hydrocarbons in the gaseous, liquid, or amorphous solid state.† These bands are assigned to the strongest CH_2-rocking vibration.

In Fig. 13 the square of the strongest frequency is plotted versus the function $(1 + \cos[N\pi/(N+1)])$. For $N < 6$ the points for *n*-hydrocarbons fall approximately on a straight line. For higher N values the curve tapers off the straight line and becomes horizontal. The reason for this is as follows: our theory holds only for a *linear* set of oscillators. An amorphous chain of a long normal hydrocarbon is not a linear zig-zag chain because the energy barrier for free rotation around a C—C bond is quite low. This means that segments of a normal hydrocarbon chain only about five carbon atoms long are straight, planar zig-zag chains. Shimanouchi and Mizushima (Ref. 26) obtained the same result from Raman spectra of the skeletal

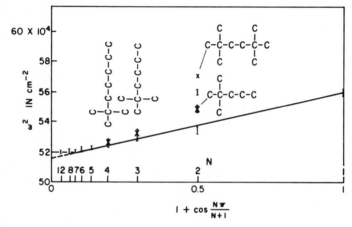

FIG. 13. Square of the observed CH_2-rocking vibrational band frequency for CH_3—$(CH_2)_N$—CH_3 plotted as a function of $1 + \cos[N\pi/(N+1)]$; $\times$ = branched hydrocarbons with N CH_2 groups.

vibrations. The different rotational isomers of polymethylene compounds have been discussed by Funck (Ref. 18).

The oscillator frequency $\omega_0 = 718$ cm^{-1} and the apparent interaction parameter $\omega' = 148$ cm^{-1} were obtained graphically from Fig. 13. The approximate value for ω' obtained from Fig. 12 is much higher (381 cm^{-1}). One would, therefore, expect the points to fall on a much steeper line in Fig. 13. This discrepancy can be explained by the fact that we do not have infinitely heavy end groups. For the extreme case of free ends the strongest band does not shift at all as a function of chain length according to Fig. 2b. The band shift for methyl end groups should, therefore, be smaller than the shift calculated wtih a theory for fixed ends. Some additional points (marked " $\times$ ") are plotted in Fig. 11 for chains with heavy end

† We consider here amorphous samples only, since in the crystalline state we would have an additional shift and a splitting of the band.

groups. As expected, they fall above the straight line for normal hydrocarbons with methyl end groups.

CH_2-*Bending Vibration*

Figure 8 shows that the transition moment of a single CH_2 group is perpendicular to the chain. A set of antiparallel vibrating dipoles is the model applicable to the CH_2-bending vibrations. The expected absorption spectrum is shown in Fig. 5b for free ends and in Fig. 5d for fixed ends.

In most observed spectra the CH_2-bending vibrational absorption appears in the 1460–1480 cm^{-1} region. Only the strongest band in the series is observed and no systematic shift of this band as a function of chain length has yet been detected. Primas and Günthard (Refs. 29, 30) calculated the position of the band series by a normal coordinate analysis. The range turns out to be 1480–1640 cm^{-1} which is in agreement with the prediction that the strongest band which is actually observed is the one at the low frequency edge of the series. It is possible that in future work the weak satellite bands of the series might be observed, especially if spectra were measured of crystalline samples at very low temperature.

CH_2-*Twisting and -Wagging Vibrations*

The assignment of CH_2-twisting and -wagging vibrations has been the subject of many papers in the literature (see Ref. 35 for a summary). Figure 9 gives the frequency ranges assigned to these vibrations by various authors. For normal hydrocarbons with N CH_2 groups, N twisting and N wagging vibrations are expected within their respective frequency intervals. Brown, Sheppard, and Simpson (Ref. 7) have given rigorous selection rules for finite planar zig-zag chains. They showed that for N even (symmetry C_{2h}), $\frac{1}{2}N$ wagging vibrations and $\frac{1}{2}N$ twisting vibrations are expected to be infrared active. For N odd, all the wagging vibrations are expected to be infrared-active, $\frac{1}{2}(N-1)$ are perpendicular bands (expected to be weak) and $\frac{1}{2}(N+1)$ are parallel bands. $\frac{1}{2}(N-1)$ twisting vibrations should be infrared-active. With the help of these selection rules and associated polarization effects, Snyder (Ref. 39) has been able to assign part of a progression band series, observed for normal hydrocarbons in the 1180–1360 cm^{-1} region to CH_2-wagging vibrations. By extrapolation he found that the series is expected to range from 1175 to 1415 cm^{-1}. The CH_2-twisting vibrations are expected to be much weaker since no dipole moment change occurs with the twisting of an individual CH_2 group.

For hydrocarbon chains with polar end groups rather strong progression band series can be observed in the 1180–1345 cm^{-1} region (Refs. 9, 10, 12, 13, 14, 16, 17, 21, 25, 29, 40, 41, 42, 44). A typical spectrum is shown in Fig. 1 of Chapter I. An extensive experimental study of these bands has been carried out by Aronovic (Ref. 2, see also Ref. 25). He as well as other workers (Refs. 11, 20, 50) have used the band series for analytical purposes (see Section 4B, Chapter I); Fig. 14 is reproduced from Aronovic's thesis. These bands might be due to CH_2-wagging vibrations corresponding to the low frequency half of the progression band series. On the average, they occur at an approximately 10 cm^{-1} lower frequency than the corresponding bands for normal hydrocarbons. For Na, Ba, and Ag salts of fatty acids there is little interference with other absorptions of the spectrum (Ref. 25). In these spectra the band series seem to be clearly confined to the 1180–1345 cm^{-1} region without any members at higher frequencies, while Susi (Ref. 40) reports that for various octadecenoic acids band series of strong

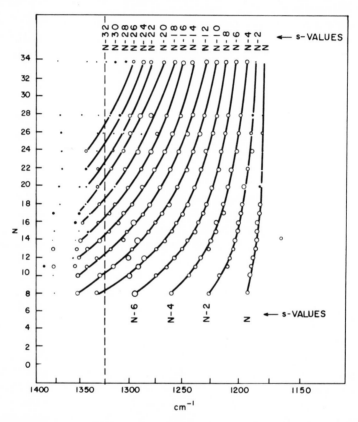

FIG. 14. Progression bands in the spectra of KBr disks of the crystalline *n*-aliphatic acids CH$_3$—(CH$_2$)$_N$—COOH. After Aronovic (Ref. 2).

bands are confined to the 1180–1310 cm^{-1} region. Meiklejohn and co-workers (Ref. 25, see also Ref. 2) found that the series contain $n/2$ (n being even) or $(n+1)/2$ (n being odd) bands, where n is the number of carbon atoms in the chain; n is equal to $N+2$ where N is the number of CH$_2$ groups. If we discard the highest frequency member of the series the bands might be assigned to the CH$_2$-twisting vibrations within the range 1180–1322 cm^{-1}. The number of bands is then $N/2$ if N is even and $(N+1)/2$ if N is odd, which agrees with the number expected from the selection rules for a chain of coupled dipoles. Equation (11b) gives the vibrational frequencies of such a chain with fixed ends. This theoretical equation is a good approximation for the observed absorption bands as shown by Fig. 15 where the square of 146 observed absorption frequencies (given in Fig. 14) is plotted against $1 + \cos[s\pi/(N+1)]$, where N varies from 8 to 34 and $s = N, N-2, N-4, \ldots$. The points fall on a straight line over quite a wide range as expected from Eq. (11b). Deviations from this straight line occur at the low and high frequency ends of the progression band series. Differences may be expected for

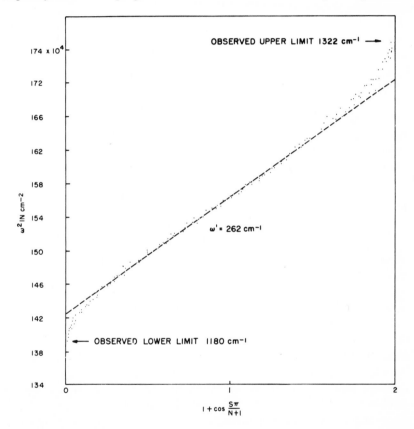

FIG. 15. Observed progression band series for n-aliphatic acids $CH_3-(-CH_2-)-_N COOH$. The diagram shows the square of the frequencies plotted against $1 + \cos[s\pi/(N+1)]$ for 146 bands shown in Fig. 14.

two main reasons. First, n-aliphatic acids do not have "infinitely heavy" end groups so that a fixed end model is not quite adequate, and second, the coupling force constants near an end group are probably somewhat different from k' in the center of the chain. Furthermore, the assignment of lines in the high frequency end of the branch is somewhat uncertain.

The observed intensity distribution in the band series is different from the predicted one for a set of coupled dipoles. This difference arises from the fact that no dipole moment change occurs during the twisting motion of a single CH_2 group, and without the inductive effect of the polar end groups, the vibrations would not be infrared-active at all. As opposed to the schematic diagrams in Fig. 5 (for infrared-active band series) the lowest frequency band is not always the strongest one in the series. This fact might be interpreted in the following way. The polar end group induces a larger dipole moment in the adjacent CH_2 groups than in the ones further away. We, therefore, would expect the highest intensity for those vibrational modes where mostly the CH_2 groups near the end of a chain oscillate. From Figs. 4a and 4b it can be seen that these are the bands near the center of the series. If on the other hand the induced dipole moments were the same in all CH_2 groups we would have a " normal " intensity distribution (Fig. 5).

CH_2-*Stretching Vibrations*

The symmetric and antisymmetric CH_2-stretching vibration bands do not show any measurable splitting or shift due to interactions of adjacent CH_2 groups. This finding is in agreement with the calculations in Section 2D (Table I) where it is shown that vibrational interaction should have a very small effect on high frequency bands.

Methyl Rocking Vibration

Normal hydrocarbons have a rather strong band in the 11μ region which has been assigned to the methyl rocking vibration caused by the methyl end groups of the chain. The same band can be observed in branched polyethylene and in copolymers of polyethylene with polypropylene, poly-1-butene, poly-1-pentene, etc. Boyd, Voter, and Bryant (Ref. 4) have found empirically a frequency dependence of the methyl band as a function of the side branch length shown in Table III.

This effect is due to the interaction of the methyl vibration with vibrations of the side branch, probably carbon-carbon skeletal vibrations. Since the specific form of interaction is not known we assume a general model as shown in Fig. 16b in order to explain the observed shifts. The large arrow corresponds to the oscillator representing the methyl vibration and the small arrows represent oscillators in the side branch. The mathematical treatment of such a system shows that no explicit solutions can be obtained but it would be relatively easy to solve the problem numerically if oscillator and coupling parameters were known. Lacking this knowledge we make the simplifying assumption that we only have an interaction between the methyl vibration and the lowest frequency component of the N side branch vibrations. This is the vibrational mode where all side branch oscillators move in phase and, therefore, have the highest influence on any adjacent oscillator. Thus the problem is reduced to the treatment of two coupled oscillators with different frequencies. Such a system with the frequencies Ω_1 and Ω_2 (before coupling) vibrates with the frequencies ω_1 and ω_2 corresponding to a symmetric and antisymmetric motion of the oscillators. If the coupling parameters Ω_1' and Ω_2' are large compared with $\Omega_1 - \Omega_2$, ω_1 and ω_2 can be expressed in the following form:

$$\omega_1^2 = \tfrac{1}{2}(\Omega_1^2 + \Omega_2^2) + \Omega_1'^2 + \Omega_2'^2 \tag{30a}$$

$$\omega_2^2 = \tfrac{1}{2}(\Omega_1^2 + \Omega_2^2) \tag{30b}$$

We now identify Ω_1 with the methyl rocking vibration and Ω_2 with the "all in phase" side branch vibration which depends on the side branch length as expressed by Eq. (29) (Ω_2 is the same as $\omega_{\text{strongest}}$ in this equation). Combining (29)

TABLE III

METHYL ROCKING VIBRATIONAL FREQUENCY AS A FUNCTION
OF SIDE BRANCH LENGTH (COMPARE FIG. 16a)

N	Absorption frequency (cm^{-1})
0	972
1	923
2	903
3	894
4	891
5	
6	
7	888
8	
9	887
10	
11	887
12	
13	887
14	
15	888

with (30a, b) we find that both frequencies ω_1 and ω_2 depend on the branch length in the following way:

$$\omega_1^2 = a \cos \frac{N\pi}{N+1} + b \tag{31}$$

where a and b are functions of oscillator and coupling force constants. A similar equation holds for ω_2.

One would expect to observe two absorption bands at ω_1 and ω_2, both shifting to lower frequencies with increasing branch length in a similar manner, but in reality only one band in the 11μ region is observed while the other one is either weak or not infrared-active at all. The square of the observed frequencies, listed in Table III, is plotted as a function of $\cos[N\pi/(N+1)]$ in Fig. 17. The points fall

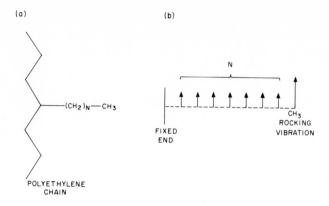

FIG. 16. Schematic representation of a side branch in polyethylene.

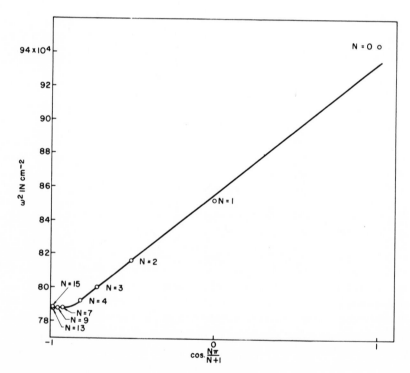

FIG. 17. Square of the CH₃-rocking vibrational frequency against $\cos[N\pi/(N+1)]$.

approximately on a straight line as expected from Eq. (31) but for long chains the points deviate from this line similar to the case shown in Fig. 13. The reason is again the fact that the side branches are no longer straight, planar zig-zag chains if N is larger than four or five.

Comparison of Normal Coordinate Treatment with the Simple Oscillator Model

Primas and Günthard (Ref. 29) have carried out a normal coordinate analysis of hydrocarbon chains with a finite length. They used force constants obtained from simple molecules and applied them to zig-zag chains. In a first approximation

TABLE IV

Frequency	CH$_2$-twisting		CH$_2$-rocking	
	calc	obs	calc	obs
Maximum	1243	1322	1109	1047
Minimum	1176	1180	717	720

they give the following two formulas for the absolute values of the frequencies of the CH$_2$-twisting and -rocking vibrations for a chain of the form R–(–CH$_2$CH$_2$–)$_{\overline{n}}$–R :

$$\Lambda^T \approx 0.9104 - 0.0384\Gamma - 0.0832\Gamma^2 + 0.0384\Gamma^3 - 0.0128\Gamma^4 \qquad (32a)$$

$$\Lambda^R \approx 0.4136 - 0.1722\Gamma + 0.0872\Gamma^2 - 0.0384\Gamma^3 + 0.0128\Gamma^4 \qquad (32b)$$

where $\Gamma = \cos[\pi k/(n+1)]$, $k = 1, 2, \ldots, n$, and $\Lambda = 0.5889 \cdot 10^{-6} \cdot \omega^2$; ω is the vibrational frequency in cm^{-1}.

Equations (32a, b) give the absolute frequency values for the expected lines in the band series. A comparison of the observed and calculated frequency intervals is given in Table IV. The agrement for the CH$_2$-rocking vibrations is quite good but there is a discrepancy for the upper limit of the twisting vibration range, which is probably due to the interference of twisting and wagging vibrational bands.

Individual lines of the band series are not observed at the frequencies calculated by Eq. (32a) for the twisting vibrations. The situation is somewhat better for the rocking vibrations where the agreement is at least semiquantitative as shown in Fig. 18 where the square of the observed frequencies for C$_{24}$H$_{50}$ (Fig. 10a) is plotted against Λ^R as calculated by Eq. (32b). For a quantitative correlation of observed and calculated values the points should fall on the solid straight line but instead they fall on a slightly curved line with a somewhat different slope. Even though the agreement is not perfect, we should not underestimate the usefulness of a normal coordinate analysis. Such an analysis is the only way to calculate the *absolute value* of the vibrational frequencies based on reasonably well-known force constants of simple molecules, while the coupled oscillator model can only describe the behavior of the band series in a relative manner.

Theimer (Refs. 45, 46, 47) and Primas and Günthard (Refs. 31, 32) have also made attempts to calculate intensities with a normal coordinate treatment but their formulas are very complex and have not yet been compared with observed spectra in a quantitative way. Furthermore, their calculations are restricted to hydrocarbons, while the oscillator model can be applied to any chain molecule.

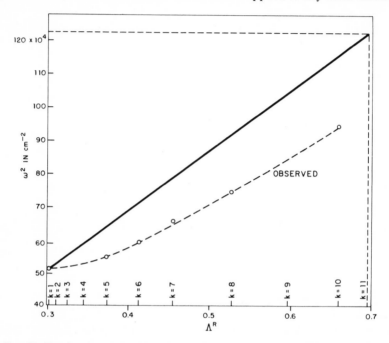

FIG. 18. Band series of the CH_2-rocking vibrations in $C_{24}H_{50}$. The square of the frequency is plotted against the function Λ^R (Eq. 32b).

C. Carbonyl Absorption Band in 1-Nylon Chains

A tentative structure of a 1-nylon chain is shown in Fig. 1a. The interaction of carbonyl stretching vibrations has briefly been discussed in the introduction of this chapter.

The observed absorption spectrum of ethyl-1-nylon is shown in Fig. 19. A thin film of this polymer has a sharp C=O stretching absorption at about 5.88μ while a thick sample shows an asymmetric band "shaded" toward higher frequencies. This "wing" of the band can be explained in the following way. For a chain with a finite length we would expect to observe a band splitting and an absorption intensity distribution as shown in Fig. 5c. For a polymer with a wide molecular weight distribution the spectrum should consist of a composite of band series resulting in an absorption wing where each molecular weight fraction contributes according to its chain length. The intensity of this wing is rather weak as expected from Eq. (27) and Fig. 5. It, therefore, can be observed only in a relatively thick sample.

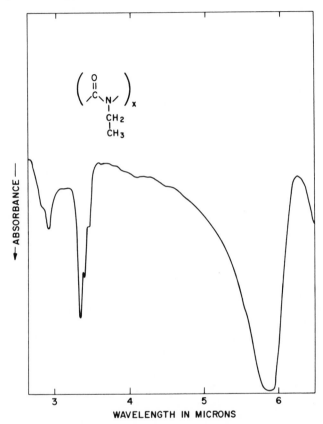

FIG. 19. Spectrum of ethyl-1-nylon in the 3–6 μ region.

D. OUTLOOK

It should be possible to apply the coupled oscillator model in the present form to a variety of chain molecules to describe absorption band shapes, shifts, and splittings. But it might be necessary to refine the model and consider a more general arrangement of oscillators along a linear axis in order to include helical polymers.

REFERENCES

1. Abrahamsson, S., *Arkiv Kemi* **14**, 65 (1959).
2. Aronovic, S. M., Ph.D. Thesis, University of Wisconsin, 1957.
3. Born, M., and von Kármán, T., *Physik. Z.* **13**, 297 (1912).

4. Boyd, D. R. J., Voter, R. C., and Bryant, W. M. D., Paper presented at 132nd meeting of the American Chemical Society, New York, 1957. Abstracts p. 8T.

5. Brillouin, L., "Wave Propagation in Periodic Structures," 2nd ed. Dover, New York, 1953.

6. Brown, J. K., and Sheppard, N., *Trans. Faraday Soc.* **50**, 535 (1954).

7. Brown, J. K., Sheppard, N., and Simpson, D. M., *Discussions Faraday Soc.* **9**, 261 (1950).

8. Brown, J. K., Sheppard, N., and Simpson, D. M., *Phil. Trans. Roy. Soc.* **A247**, 35 (1954).

9. Chapman, D., *J. Chem. Soc.* p. 4489 (1957).

10. Chapman, D., *J. Chem. Soc.* p. 784 (1958).

11. Childers, E., and Strutters, G. W., *Anal. Chem.* **27**, 737 (1955).

12. Cole, A. R. H., and Jones, R. N., *J. Opt. Soc. Am.* **42**, 348 (1952).

13. Corish, P. J., and Chapman, D., *J. Chem. Soc.* p. 1746 (1957).

14. Corish, P. J., and Davison, W. H. T., *J. Chem. Soc.* p. 927 (1958).

15. Deeds, W. E., Ph.D. Thesis, Ohio State University, 1951.

16. Ferguson, E. E., *J. Chem. Phys.* **24**, 1115 (1956).

17. Fischmeister, I., and Nilsson, K., *Arkiv Kemi* **16**, 347 (1961).

18. Funck, E., *Z. Elektrochem.* **62**, 901 (1958).

19. Gotlib, Yu. Ya., *Opt. Spectry. (U.S.S.R.) (English Transl.)* **7**, 191 (1959).

20. Jahn, A. S., and Susi, H., *J. Phys. Chem.* **64**, 953 (1960).

21. Jones, R. N., McKay, A. F., and Sinclair, R. G., *J. Am. Chem. Soc.* **74**, 2575 (1952).

22. Lagrange, J. L., "Méchanique analytique," 4th ed., Vol. 1, p. 405. 1888. (First edition, p. 317. Veuve Desaint, Paris, 1788.)

23. Martin, J. M., Johnston, R. W. B., and O'Neal, M. J., *Spectrochim. Acta* **12**, 12 (1958).

24. McMurry, H. L., and Thornton, V., *Anal. Chem.* **24**, 318 (1952).

25. Meiklejohn, R. A., Meyer, R. J., Aronovic, S. M., Schuette, H. A., and Meloch, V. M., *Anal. Chem.* **29**, 329 (1957).

26. Mizushima, S., "Structure of Molecules and Internal Rotation." Academic Press, New York, 1954.

27. Parodi, M., *Mem. Sci. Phys.* (Paris) **47**, 63 (1944).

28. Pimentel, G. C., and Klemperer, W. A., *J. Chem. Phys.* **23**, 376 (1955).

29. Primas, H., and Günthard, H. H., *Helv. Chim. Acta* **36**, 1659 (1953).

30. Primas, H., and Günthard, H. H., *Helv. Chim. Acta* **36**, 1791 (1953).

31. Primas, H., and Günthard, H. H., *Helv. Chim. Acta* **38**, 1254 (1955).

32. Primas, H., and Günthard, H. H., *Helv. Chim. Acta* **39**, 1182 (1956).

33. Routh, E. J., "Dynamics of a System of Rigid Bodies," 6th ed. Dover, New York, 1955.

34. Rutherford, D. E., *Proc. Roy. Soc. Edinburgh* **A62**, 229 (1947).

35. Sheppard, N., *Advan. Spectry.* **1**, 288 (1959).

36. Sheppard, N., and Simpson, D. M., *Quart. Rev. (London)* **6**, 1 (1952).

37. Smith, A. E., *J. Chem. Phys.* **21**, 2229 (1953).

38. Snyder, R. G., *J. Chem. Phys.* **27**, 969 (1957).

39. Snyder, R. G., *J. Mol. Spectry.* **4**, 411 (1960); **7**, 116, (1961).

40. Susi, H., *Anal. Chem.* **31**, 959 (1959).

41. Susi, H., *J. Am. Chem. Soc.* **81**, 1535 (1959).

42. Susi, H., and Smith, A. M., *J. Am. Oil Chemists' Soc.* **37**, 431 (1960).

43. Sutherland, G. B. B. M., and Sheppard, N., *Nature* **159**, 739 (1947).

44. Sydow, E. von, *Acta Chem. Scand.* **9**, 1119 (1955).

45. Theimer, O. H., *Z. Naturforsch.* **11a**, 883 (1956).

46. Theimer, O. H., *J. Chem. Phys.* **27**, 408 (1957).

47. Theimer, O. H.,*J. Chem. Phys.* **27**, 1041 (1957).
48. Tschamler, H.,*J. Chem. Phys.* **22**, 1845 (1954).
49. Vidro, L. I., and Volkenstein, M. V., *Dokl. Akad. Nauk S.S.S.R.* **85**, 1243 (1952).
50. Wenzel, F., Schiedt, U., and Breusch, F. L., *Z. Naturforsch.* **12b**, 71 (1951).
51. Whitcomb, S. E., Nielsen, H. H., and Thomas, L. H.,*J. Chem. Phys.* **8**, 143 (1940).
52. Zbinden, R.,*J. Mol. Spectry.* **3**, 653 (1959).
53. "Infrared Spectral Data." Am. Petrol. Inst., Pittsburgh, Pennsylvania, 1957.

—V—

Orientation Measurements

1. Introduction

In Chapter I we have briefly discussed the effects of molecular orientation on the spectrum of a polymer sample. This problem will now be considered in more detail and treated in a quantitative mathematical way as far as is possible at the present time.

Two aspects of orientation measurements are of practical as well as theoretical importance. First, it is possible to determine the physical structure of a polymer sample such as a fiber or a film. The type and degree of orientation of polymer chains and crystallites can be measured. Second, it is possible to obtain information about the chemical as well as the geometrical structure of the individual polymer chains. For example, in some cases one can determine the angle of a certain chemical group (its transition moment) with respect to the chain axis. The specific methods will be described in the following sections.

Crystallinity and orientation are often closely related in a polymer sample and their effects on the spectrum are usually interdependent, but we will make an attempt to separate the two effects in order to simplify the treatments.

First, we will discuss some experimental problems with respect to sample preparation and orientation measurements. Then we will analyze various types of orientation and calculate the expected dichroic ratios.

2. Experimental Techniques and Dichroic Ratio Measurements

A. Sample Preparation

In many cases it is possible to apply conventional sample preparation methods (Ref. 5) to polymers. Nujol mulls or KBr pellets can be prepared for most polymers although it might be difficult to grind a very tough one. In such a case it is necessary to work at liquid nitrogen or even lower temperatures in order to make the sample brittle. Spectra of some polymers can also be obtained as solutions, but polymer solvents with wide " windows " in the infrared region are relatively rare.

The form of a polymer most convenient for infrared analysis is a film, which might be prepared by melt pressing or casting from a solution.† The film samples

---

† There are various other methods for preparing films. Some are only applicable to specific polymers, for example, some films can be obtained by polymerizing a monomer between two glass plates.

should have a minimum size in order to cover the infrared light beam. This minimum size depends upon the optical system of the spectrophotometer. Approximate values are listed in Table I. The thickness of the sample should be in the range of about 1 to 200μ depending on the extinction coefficient of the absorption bands to be analyzed.

Many natural and synthetic polymers are available in fiber form. The microspectrometer is particularly useful to obtain spectra of such samples directly without any further sample preparation.

Some fibers and films can be drawn under suitable conditions to orient the polymer chains along the draw direction. The polarized spectrum of such oriented samples contains more information than the spectrum of randomly oriented chains.

TABLE I

APPROXIMATE MINIMUM SAMPLE DIMENSIONS PERPENDICULAR TO THE LIGHT BEAM IN MILLIMETERS

Instrument	Parallel to the monochromator slit	Perpendicular to the monochromator slit
1. Ordinary spectrophotometer such as Perkin-Elmer model 21	15	4
2. Same as 1, but sample placed directly in front of entrance slit	10	2
3. Same as 1, but combined with a beam condensing system	2	0.5
4. IR microscope	0.3–1	0.01–0.03

Sample preparation and orientation methods are usually different for different polymers and the experimenter will have to find suitable conditions for each individual case. The sample preparation will also depend on the type of problem to be solved. If, for example, one wants to know the degree of orientation in a rolled film, a sample suitable for infrared analysis should be obtained without changing this orientation, but if one is interested in the steric configuration and chemical structure of the chain a change in degree of orientation does not matter.

B. POLARIZER AND INSTRUMENT POLARIZATION

A variety of polarizers have been designed for infrared radiation (Refs. 21, 22, 41, 48), but the one most commonly used in recent years consists of about six silver chloride plates (Fig. 1) placed in the light beam at Brewster's angle α (Refs. 44, 60) which is determined by the equation

$$\tan \alpha = n \tag{1}$$

where n is the refractive index of silver chloride. n and, therefore, α are wavelength dependent but for practical purposes an average value of α is used for a wide wavelength range. For silver chloride α is about 63.5°. The degree of polarization for such a polarizer is of the order 95–98%. Bird and Shurcliff (Ref. 9) have designed an improved version of this polarizer using six plates that are slightly wedge shaped and mounted in a "fanned" manner instead of parallel. They also pointed out that the degree of polarization can be increased if the plates are placed at an angle somewhat larger than Brewster's angle, namely, at approximately 68°, with only a slight loss in light intensity. The deficiency in polarization of a few percent leads to errors in dichroic ratios. In Section 2D we will discuss how such errors can be corrected.

An effect which will have to be considered is the partial polarization in the spectrophotometer itself. This polarization is caused by reflections at various mirrors and especially at the prism faces. The degree of instrument polarization

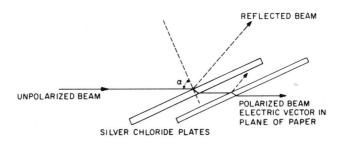

FIG. 1. Principle of silver chloride plate polarizer, two plates are shown.

(electric vector parallel to the slit) is usually about 15 to 20% for a single pass and about twice that amount for a double pass instrument. The spectrum of an oriented sample (e.g., a drawn film strip) taken with unpolarized light will, therefore, not be the same whether the draw direction is parallel or perpendicular to the slit. To obtain an "unpolarized" spectrum one can place the sample in the light beam so that the draw direction forms an angle of 45° with the slit. This same compensation method can also be used to facilitate dichroic measurements. The oriented sample again is placed at an angle of +45° to the slit while the spectra are recorded with the plane of polarization at +45° and −45°, respectively. The sample can, of course, only be placed in this position if it is still wide enough to cover the whole cross section of the incident light beam after the rotation. This is not always possible with long narrow samples such as fibers.

Finally, we ought to point out that whenever possible the sample should not be placed between polarizer and monochromator (which is a partial polarizer by itself) since the plane of polarization might be rotated by the sample's birefringence causing an error in the dichroic ratio (Refs. 9, 33).

C. The Infrared Microspectrometer

The infrared microscope is an attachment to a spectrometer with which it is possible to obtain spectra of extremely small samples. This is of importance in the field of biology where often only a few micrograms of a compound are available. It should also turn out to be a powerful tool to analyze small samples collected from a gas chromatographic column. In this chapter we are particularly interested in using the infrared microscope in combination with polarized radiation for the analysis of small single crystals and oriented fibrous samples of natural as well as synthetic origin (Refs. 10, 16, 17, 40).

We will now briefly describe the principle of the microscope and discuss some points which are different from ordinary infrared spectroscopy.

Most spectrometers require a minimum sample size of the order of 2×12 mm in order to cover the light beam at its smallest cross section. Spectra of smaller samples can be obtained but only with an appreciable loss of energy. The use of a microscope attachment can solve this difficulty, and samples about two orders of magnitude smaller can easily be run. All the microscopes described in the literature (Refs. 3, 4, 11, 12, 20, 27, 30, 49, 58) work alike in principle: A very small image of the infrared radiation is produced by an optical system with a high numerical aperture, and the sample to be analyzed is placed at the position of this image. In this fashion spectra of samples as small as about 0.01×0.6 mm can be obtained. In some cases these dimensions might be even smaller depending on the wavelength of the radiation, as discussed by Fraser (Ref. 25). Various optical systems have been designed but the Schwarzschild-type reflection objective is the one most commonly used today. Two more modest beam condensing systems have been described in the literature (Refs. 2, 23). One is a reflection type system and the other one consists of two silver chloride lenses. Both these systems are extremely simple and, therefore, very useful. They reduce the image by a factor of 5 to 10 which is sufficient for a number of problems.

Most of the earlier microscope attachments were placed in the undispersed beam of the spectrometer so that all the energy from the light source had to pass through a very small sample. In some cases this can cause melting or degradation of the sample, especially if a carbon or zirconium arc (Refs. 18, 51) is used as a light source. To avoid this effect Coates, Offner, and Siegler (Ref. 19) have described an instrument with the microscope attachment placed in the dispersed beam where only little radiation passes through the sample at any one moment when a spectrum is recorded. Such an instrument is commercially available from the Perkin-Elmer Corporation and its optical layout is shown in Fig. 2. The monochromatic light leaves the exit slit of the monochromator, passes through the condenser, and forms an image of the slit about eight times reduced at the position of the sample. The objective with a magnification of about $25 \times$ forms a new image of the slit and the sample at the plane of the diaphragm. Finally the light is focused on the detector with a third Schwarzschild condenser. The diagonal mirror M is used for a convenient layout of the instrument and, therefore, is of no principal importance. More details can be found in Ref. 19.

Let us now discuss some problems in obtaining spectra of very narrow samples such as fibers. If a fiber is placed at the first image of the exit slit and this image is wider than the sample, only part of the radiation will pass through the fiber. The

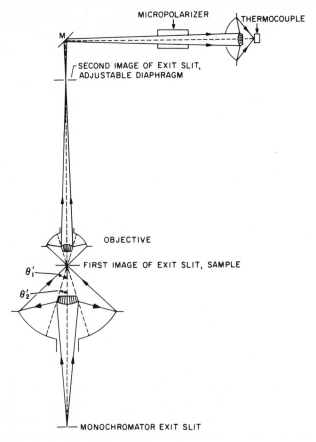

FIG. 2. Infrared microscope (model of Perkin-Elmer Corporation except for micro-polarizer).

other part, that by-passed the sample, can be cut off with the adjustable diaphragm at the second image so that no " false radiation " can reach the detector. But there are still other sources for false radiation. The sample size is often comparable to the wavelength of the radiation which, therefore, is partially diffracted around the sample without being absorbed. Furthermore, some false radiation can reach the detector because of an imperfect optical system. False radiation causes the apparent intensity of an absorption band to be too small. This effect has, therefore, been called "spectral dilution" (Refs. 13, 19). It can be reduced by narrowing the diaphragm and monochromator slit width but this reduces the total amount of energy reaching the detector. The energy can be increased again by using a monochromator with a low dispersion (e.g., with a CsBr prism in the $2-8\mu$ region) and sacrificing some resolution. To summarize, with a high dispersion monochromator (high resolution) the slit is wide and an absorption band weakened due to spectral dilution; with low dispersion, on the other hand, the bands are weakened because of poor resolution. Therefore, there exists for each individual case an "optimum dispersion" which leads to a maximum apparent absorption band intensity. This is opposed to ordinary infrared spectroscopy where the best spectrum (smallest deformation of absorption band) is obtained with the highest resolution monochromator. To illustrate this effect some measurements by Bohn (Ref. 13) are

TABLE II

OPTICAL DENSITY FOR THREE ABSORPTION BANDS OF A POLYACRYLONITRILE FIBER

Prism	3.4μ CH_2-stretching band	4.4μ —C≡N stretching band	6.9μ CH_2-bending band
Cesium bromide	0.106	0.106	0.267
Sodium chloride	0.133	0.232	0.300[a]
Sodium chloride and C arc	0.149	0.325	0.420
Calcium fluoride	0.180	0.374[a]	0.330[a]

[a] Image of slit is wider than sample.

listed in Table II. The optical densities are listed for three different absorption bands of polyacrylonitrile. All the measurements were carried out at a constant diaphragm width and a constant signal to noise ratio. The cesium bromide prism has the lowest and the calcium fluoride prism the highest dispersion. For the 3.4 and 4.4μ bands the apparent band intensity is highest for the calcium fluoride prism, but this is no longer true for the 6.9μ band where the wider slit width required for this prism and the large wavelength cause appreciable spectral dilution.

The microscope in combination with a polarizer is a very powerful tool for orientation measurements on a narrow filament or a small single crystal. The polarizer is usually placed in the undispersed beam but in combination with a high intensity light source (Refs. 18, 51). This is not practical since the silver chloride plates darken very rapidly due to photodegradation. Bohn (Ref. 13) has, therefore, designed a micropolarizer to fit in the dispersed beam between the diaphragm and the thermocouple condenser (Fig. 2). This polarizer is equivalent to or somewhat better than the conventional one since particularly transparent sections of silver chloride sheets can be selected for the micropolarizer.

The use of the microscope introduces some errors in dichroic measurements due to the spectral dilution and the highly converging beam at the position of the sample. These errors will be discussed in the following subsections.

D. Sources for Errors in Dichroic Ratio Measurements

The dichroic ratio R of an absorption band is defined by the ratio of its integrated intensities measured with light polarized parallel and perpendicular to a given direction (e.g., direction of drawing) in the sample. In most practical cases one measures the ratio of the peak intensities of the absorption maximum of the band which is identical with R if the band shape of the parallel and perpendicular component is the same. This is true in many practical cases where both components can be approximated, for example, by a Lorentz-type equation (Ref. 32). We, therefore, will consider only peak intensities in future discussions on dichroic measurements, but we should keep in mind that in some cases the ratio of peak intensities can be considered only an approximation to the true dichroic ratio.

The determination of dichroic ratios encounters the same difficulties as ordinary intensity measurements as soon as various absorption bands overlap where it becomes uncertain how to measure the background intensity for an individual band. The error introduced in dichroic measurements is not quite as serious as in the determination of a single band intensity since the errors in the parallel and perpendicular components might be similar and cancel out to some extent on formation of the dichroic ratio. There are a number of other effects which can falsify dichroic ratio measurements. Some of these effects will now be discussed.

Polarizer Efficiency

The polarizer efficiency is usually about 95 to 99% as discussed in subsection B. This will cause the measured dichroic ratio to be closer to unity than the true one. The error will decrease with increasing polarizer efficiency. A quantitative discussion of this problem will now be given.

Let us consider a dichroic absorption band with a weak (w) and strong (s) component in the two polarization directions respectively, and define the true dichroic ratio R as the ratio of the two optical densities A:

$$R = \frac{A_s}{A_w} = \log \frac{I_0}{I_s} \bigg/ \log \frac{I_0}{I_w} \tag{2}$$

A is defined by Beer's law:

$$\log \frac{I_0}{I} = \alpha d = A \tag{2a}$$

α is the extinction coefficient and d is the sample thickness. The intensities I for an ideal polarizer are shown in Fig. 3. A real polarizer has some leakage and in the parallel position, for example, the fraction ϵ of the perpendicular polarized light leaks through (and vice versa). The true transmission intensities T are then given by

$$T_0 = I_0 + \epsilon I_0$$

$$T_w = I_w + \epsilon I_s \qquad (3)$$

$$T_s = I_s + \epsilon I_w$$

and the observed dichroic ratio R_{obs} is

$$R_{\text{obs}} = \log \frac{T_0}{T_s} \bigg/ \log \frac{T_0}{T_w} = \log \frac{I_0(1+\epsilon)}{I_s + \epsilon I_w} \bigg/ \log \frac{I_0(1+\epsilon)}{I_w + \epsilon I_s} \qquad (4)$$

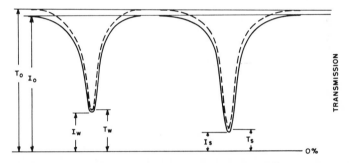

FIG. 3. Influence of imperfect polarizer. w and s are the weak and strong components, respectively. KEY: ———: perfect polarizer; – – –: imperfect polarizer.

combining (4) with (2) we obtain R_{obs} as a function of the true R and the optical density A_s of the stronger of the two components:

$$R_{\text{obs}} = R \frac{1 + \dfrac{1}{A_s} \log \dfrac{1+\epsilon}{1 + \epsilon(I_w/I_s)}}{1 + \dfrac{R}{A_s} \log \dfrac{1+\epsilon}{1 + \epsilon(I_s/I_w)}} = R \frac{1 + \dfrac{1}{A_s} \log \dfrac{1+\epsilon}{1 + \epsilon \exp\{-2.303 A_s(R^{-1}-1)\}}}{1 + \dfrac{R}{A_s} \log \dfrac{1+\epsilon}{1 + \epsilon \exp\{2.303 A_s(R^{-1}-1)\}}} \qquad (5)$$

In most cases ϵ is small compared to 1 and (5) can be somewhat simplified by a suitable approximation, but this will not be discussed in further detail. The results of Eq. (5) are shown graphically in Fig. 4 where R_{obs} is plotted against R for various optical densities A_s of the strong component. A polarizer efficiency of 98% was assured ($\epsilon = 0.02$). This plot can be used directly for correcting the observed dichroic ratio measurement. For practical purposes $A_{s,\text{obs}}$ can be used instead of A_s since this introduces only a second order error.

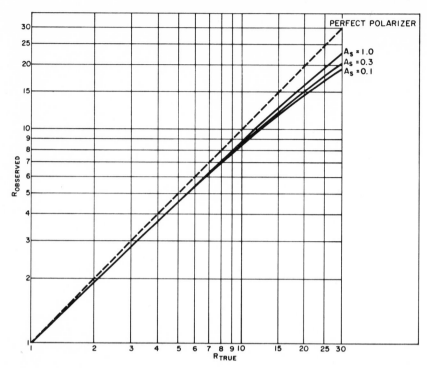

FIG. 4. Observed against true dichroic ratio for various optical densities of the stronger component A_s. Polarizer efficiency = 98%.

Insufficient Resolution of the Spectrometer

If the resolution of the monochromator is insufficient an absorption band is deformed and the observed peak intensity is smaller than the true one. Both components of a dichroic band will be weakened and the intensity ratio will not be affected in a first approximation, but since the intensity reduction is not quite the same for the strong and the weak component, the second order effect on the dichroic ratio might be appreciable. Ramsay (Ref. 50) has discussed the influence of the instrument resolving power on the band intensity, assuming a Lorentz-type band shape. He has calculated the expected intensity reduction as a function of the observed intensity and the resolving power of the monochromator. His results are listed in Table III.

The resolution is expressed by a dimensionless parameter $S/\Delta\nu_{1/2}$ where S is the spectral slit width and $\Delta\nu_{1/2}$ is the observed absorption band half width, both expressed in cm^{-1}. The following example shows how Table III can be used directly to calculate the real dichroic ratio from the observed one. If $S/\Delta\nu_{1/2} = 0.65$ and $R_{\text{obs}} = 8$ with the two components having an optical density

TABLE III[a]

$S/\Delta\nu_{1/2}$

A_{obs}	0.00	0.05	0.10	0.15	0.20	0.25	0.30	0.35	0.40	0.45	0.50	0.55	0.60	0.65
0.87	1.00	1.00	1.01	1.02	1.03	1.04	1.06	1.09	1.13	1.18	1.24	1.32	1.44	1.61
0.78	1.00	1.01	1.01	1.02	1.03	1.04	1.06	1.09	1.13	1.18	1.24	1.32	1.44	1.59
0.69	1.00	1.01	1.01	1.02	1.03	1.04	1.06	1.09	1.13	1.18	1.24	1.32	1.43	1.58
0.61	1.00	1.01	1.01	1.02	1.03	1.04	1.06	1.09	1.13	1.18	1.24	1.32	1.43	1.57
0.52	1.00	1.01	1.01	1.02	1.03	1.04	1.06	1.09	1.13	1.17	1.24	1.31	1.42	1.56
0.43	1.00	1.01	1.01	1.02	1.03	1.04	1.06	1.09	1.13	1.17	1.24	1.31	1.42	1.55
0.35	1.00	1.01	1.01	1.02	1.03	1.04	1.06	1.09	1.13	1.17	1.24	1.31	1.41	1.54
0.26	1.00	1.01	1.01	1.02	1.03	1.04	1.06	1.09	1.13	1.17	1.24	1.31	1.41	1.53
0.17	1.00	1.01	1.01	1.02	1.03	1.04	1.06	1.09	1.13	1.17	1.23	1.30	1.40	1.53
0.09	1.00	1.01	1.01	1.02	1.03	1.04	1.06	1.09	1.13	1.17	1.23	1.30	1.40	1.52

[a] This table (Ref. 50) contains the factor by which the observed optical density A_{obs} has to be multiplied to obtain the true value as a function of A_{obs} and $S/\Delta\nu_{1/2}$. The spectral slit width S is the half width of the curve that would be observed for a monochromatic absorption line. $\Delta\nu_{1/2}$ is the observed absorption band half width.

of 0.8 and 0.1, the true optical densities are 1.27 and 0.152, respectively, and the true dichroic ratio becomes 8.36.

False Radiation Due to Spectral Dilution and Scattered Light

In Section 2C we have shown that in an infrared microscope false radiation of the proper wavelength can reach the detector due to diffraction effects at the sample and imperfections of the optical system. This will cause the observed absorption bands to be weaker than the true ones (spectral dilution). False radiation due to scattered light of a shorter wavelength has the same effect on the band

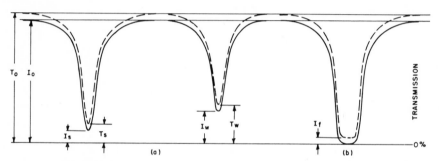

FIG. 5. Influence of false radiation on absorption bands. KEY: ———: no false radiation; ---: intensity of false radiation $= I_f$, all absorption bands are shifted by this amount.

intensity. If the intensity of the false radiation is I_f all absorption bands are shifted by this amount (Fig. 5) and the observed transmissions will have to be converted in the following way:

$$I_0 = T_0 - I_f$$
$$I_s = T_s - I_f \tag{6}$$
$$I_w = T_w - I_f$$

The true dichroic ratio can then be simply obtained by Eq. (2). Various methods are known for determining I_f. Scattered light at a wavelength λ is usually measured by placing a filter in the beam which is totally absorbing at λ but transmits shorter wavelength light causing the scattered radiation. A CaF_2 plate, for example, is a convenient filter to use for the 10–15μ region. The false radiation I_f in an infrared microscope can be determined by placing a totally absorbing object instead of the sample into the light beam. This object should have the same shape as the sample. For example, a fiber can be replaced by a wire of the same dimensions (Ref. 13). In another method of determining the false radiation one can use the sample itself as a "blackout" object if it has an extremely strong absorption band in the wavelength region of interest. Such a band can be used to determine I_f as shown in Fig. 5b.

Beam Convergence

Almost all infrared absorption measurements are carried out in a convergent beam. In such a beam the electric vector has a component parallel to the optical axis (for off-axis rays) which causes some error in polarization measurements. This error is negligibly small for ordinary instruments where a relatively large sample is available and the angle of the radiation cone is only a few degrees. But in microscopes with a high numerical aperture an appreciable error in dichroic measurements is introduced by this effect. Fraser (Ref. 26) has given a formula for the true dichroic ratio of an axial oriented sample as a function of the microscope geometry. He obtained

$$R = \log \frac{T_0}{T_\parallel + M(T_\perp - T_\parallel)} \bigg/ \log \frac{T_0}{T_\perp} \tag{7}$$

where $M = 1 - \frac{3}{2}(\cos\theta_1 + \cos\theta_2)/(\cos^2\theta_1 + \cos^2\theta_2 + \cos\theta_1\cos\theta_2)$. θ_1 and θ_2 are the semiangles of the radiation cone in the sample. T_0 is the reference intensity $T_\parallel$ and $T_\perp$ are the transmissions at the absorption band peak for radiation polarized parallel and perpendicular to the slit, respectively. The uncorrected observed dichroic ratio is simply given by

$$R_{\text{obs}} = \log \frac{T_0}{T_\parallel} \bigg/ \log \frac{T_0}{T_\perp} \tag{8}$$

The semiangles θ_1 and θ_2 are determined by the geometry and the refractive index n of the sample. For a filmlike sample they are given by

$$\sin\theta_1 = \frac{\sin\theta_1'}{n}, \qquad \sin\theta_2 = \frac{\sin\theta_2'}{n} \tag{9}$$

where θ_1' and θ_2' are the semiangles of the radiation cone in the microscope as shown in Fig. 2. For samples with other cross sections, for example, round fibers, Eqs. (9) are not correct for all the rays in the radiation beam.

Equation (7) can be used to calculate the true dichroic ratio R, but the refractive index of the sample at the absorption band is usually not known and one has to rely on an approximate value for n. This introduces an uncertainty in the corrected dichroic ratio. It, therefore, is useful to determine the correction independently by an experimental method where a large film sample is analyzed in an ordinary spectrometer with a very low aperture as well as in the microscope. The following example will illustrate this problem.

Example

For a microscope as shown in Fig. 2 the semiangles are approximately $\theta_1' = 48°$, $\theta_2' = 18°$ and for a refractive index of 1.4 one obtains from (9): $\theta_1 = 32°$, $\theta_2 = 13°$ which leads to a value of $M = -0.097$. The dichroic behavior of the —C≡N stretching vibrational band (2230 cm^{-1} region) of a drawn polyacrylonitrile film was analyzed in three ways: (I) The film was placed in the microscope with the draw direction parallel to the spectrometer slit, (II) with the draw direction perpendicular to the slit, and (III) the film was placed in an ordinary spectrometer (where the beam convergence is negligible†), with the draw direction at 45° to the slit as explained in Section 2B. In all three cases the dichroic ratio was determined by measuring the transmission values T with the polarizer parallel and perpendicular to the draw direction. The "observed" and corrected dichroic ratios were computed according to Eqs. (8) and (7), respectively. The results are listed in Table IV. In case III a dichroic ratio of 3.77 was obtained which is somewhat larger than the corrected values in Table IV.

TABLE IV

	T_0	$T_\parallel$ polarizer parallel to the slit	$T_\perp$ polarizer perpendicular to the slit	Eq. (8) R_{obs}	$\dfrac{1}{R_{obs}}$	Eq. (7) Corrected R	$\dfrac{1}{R}$
I. Draw direction parallel to the slit	0.595	0.410	0.188	0.323	3.09	0.278	3.60
II. Draw direction perpendicular to the slit	0.632	0.222	0.459	3.27		3.61	

In practice corrections to the observed dichroic ratios in a high aperture microscope can be applied in either a fully empirical or in a semiempirical way. It is possible to prepare a number of test samples with absorption bands covering a wide range of dichroic ratios and intensities. These samples can then be analyzed in a microscope as well as in an ordinary spectrometer in order to determine the corrections for various conditions. On the other hand, this correction can be determined for just one case from which the value of the parameter M in (7) is calculated. Any further corrections to other samples can then be obtained from this equation.

Concluding Remark

In a practical case all the sources mentioned above might cause errors in dichroic measurements, and the question arises in which order the various corrections should be applied. In most cases they are small (of the order of a few percent) and the order does not matter. But if one error is relatively large, for

† For an extreme case with a dichroic ratio of 15 and a parallel band of an optical density of 0.7, the correction due to convergence would only be about 2% which is usually within the experimental error.

example, the error due to false radiation, the dichroic ratio should first be corrected for this one.

E. Measurement of Dichroic Ratios Close to Unity

For polymer samples which are only slightly oriented the dichroic ratio R is very close to unity. In such a case the difference $R-1$ is a measure for the amount of orientation. The accuracy in determining the difference is usually quite poor since R is the ratio of two almost equal optical densities.

To measure small amounts of orientation more accurately, Tink (Ref. 56) has devised a method in which the difference $R-1$ is measured directly by using a double beam spectrophotometer. Two polarizers are used in this method. One is placed in the sample beam and one in the reference beam. The two planes of

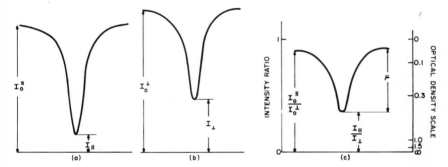

Fig. 6. Dichroic ratios close to unity. (a) Sample beam; (b) reference beam; (c) recorder, the dichroic coefficient μ is measured in optical density units (Ref. 56).

polarization are at right angles with respect to each other. The sample is placed in the spectrometer at a position where geometrically both beams are combined. With this arrangement the recorder measures the intensity ratio of the two beams, one being polarized parallel and the other one perpendicular to a given direction in the sample as shown in Fig. 6. The dichoric ratio as defined by Eq. (2) is

$$R = \log\frac{I_0^\parallel}{I_\parallel} \Big/ \log\frac{I^\circ}{I_\perp} = \frac{A_\parallel}{A_\perp} \tag{10}$$

If the chart paper on the recorder is calibrated in optical density units, the apparent band intensity μ as shown in Fig. 6c is given by

$$\mu = \log\left(\frac{I_0^\parallel}{I_0^\perp} \Big/ \frac{I_\parallel}{I_\perp}\right) = A_\parallel - A_\perp \tag{11}$$

Tink has called μ the " dichroic coefficient." From (10) and (11) we obtain

$$R = 1 + \frac{\mu}{A_\perp} \tag{12}$$

With this equation the dichroic ratio R can be calculated with a high degree of accuracy where μ is determined by the method just described and $A_\perp$ can be obtained from a conventional single beam intensity measurement.

This method for measuring low dichroism was also discussed by Stein (Ref. 54). He described a somewhat different experimental arrangement using a single beam instrument with a rotating polarizer synchronized with the shutter frequency.

In later sections we will give only correlations between various types of orientation and the dichroic ratio R, but with the help of Eq. (12) they can easily be rewritten with the dichroic coefficient μ as a variable, which might be more convenient in some practical cases.

3. Orientation and Infrared Dichroism

A. INTRODUCTORY REMARKS

In this and the following section we will discuss the mathematical problem of correlating quantitatively observed dichroic ratios with the actual structure and orientation of chain molecules. We will assume that samples are homogeneous with respect to their physical state, for example, that a sample is either fully crystalline or fully amorphous. This is, of course, a drastic simplification since actual polymers usually consist of a mixture of crystalline and amorphous fractions. In such a case the absorption spectrum is a superposition of the spectra of the two components where each of them can be treated separately (Ref. 53).

Let us now discuss the relation between orientation in a polymer sample and the dichroism of its infrared absorption bands. Consider a short portion of a polymer chain, for example, an individual amide group in a 6-6 nylon sample. Such a group has various normal modes of vibration which can be considered as characteristic group vibrations.† As an example we consider the N—H stretching motion which absorbs at about 3300 cm^{-1}. In this vibration the N and H atoms oscillate approximately along the N—H bond which becomes the direction of dipole moment change (transition moment). If we could analyze one individual amide group with radiation polarized in various directions we would observe a strong absorption band with the electric vector parallel to the N—H bond. Radiation with the electric vector perpendicular to this bond would not be absorbed at all. An actual sample of 6-6 nylon contains many amide groups oriented in various directions. The dichroic behavior of such a sample is determined by a superposition of the contributions from each individual group. Suppose we know the geometrical orientation of all the groups in a partially oriented sample, then it is possible to calculate the expected relative absorption band intensity for any angle of incidence and any polarization direction.

† In many cases an absorption band cannot be identified with an individual group in the chain but rather with a chain segment containing a number of groups with vibrational interaction (see chapter IV). In this case the orientation of the long segment rather than the individual group determines the dichroic behavior of the absorption band.

In practice, the exact orientation of the individual transition moments within a sample is not known, but the dichroic properties of a sample can be obtained experimentally. The problem is, therefore, to determine the orientation of chains or chain segments from the observed dichroic data. This can be done in an indirect way only. One will have to assume a model for a certain orientation distribution, calculate the expected dichroic properties and compare them with the observed values. In this fashion one can determine which model fits the experimental data best. In Section 4 a number of such models will be discussed.

A word of caution is in order. For a given molecular model one often pretends to know the direction of the vibrational transition moments relative to the direction of the chemical bonds. This is probably justified for some highly localized vibrations such as C—H, S—H, N—H stretching vibrations where the transition moment is expected to be along the chemical bond. For symmetric and antisymmetric CH_2-stretching vibrations as well as some other vibrations involving the CH_2 group, the transition moment is also fairly well defined (see Fig. 8, Chapter IV). But the transition moment of the C=O stretching vibration for an amide group, for example, forms an angle of about 10 to 20° with the carbonyl bond and similar deviations are expected for many other vibrations (Refs. 1, 15, 52).

In the following subsections we will describe in general terms the relation between orientation and dichroic properties.

B. Coordinate Transformations

Consider two coordinate systems (Fig. 7). In one of them the axes a, b, and c are conveniently adapted to an individual polymer chain or chain segment where the c-axis might be chosen parallel to the chain axis. The other system with the axes x, y, and z is adapted to the sample shape. For films the xy-plane may coincide with the plane of the film, and for drawn samples the z-axis is usually chosen to be parallel to the draw direction. Once the sample is fixed in space, for example, in a spectrometer, the x, y, z system is also fixed, while the a, b, c system can assume various directions according to the orientation of the individual chain segments or transition moments, respectively.

Suppose a transition moment vector **M** (e.g. the one related to the amide N—H stretching vibration, discussed above) has the three components M_a, M_b, M_c in the chain coordinate system, then the components M_x, M_y, M_z are given by a transformation of the following form† :

$$(M_x, M_y, M_z) = \begin{pmatrix} T_{11} & T_{12} & T_{13} \\ T_{21} & T_{22} & T_{23} \\ T_{31} & T_{32} & T_{33} \end{pmatrix} \begin{pmatrix} M_a \\ M_b \\ M_c \end{pmatrix} \tag{13}$$

† This is a matrix vector multiplication where, for example, the x-component M_x of the transition moment **M** is given by

$$M_x = T_{11}M_a + T_{12}M_b + T_{13}M_c$$

or abbreviated:

$$\mathbf{M}_{x,y,z} = (T)\,\mathbf{M}_{a,b,c} \tag{13a}$$

(T) is a transformation matrix and depends upon the orientation of an individual chain as well as the choice of the coordinate system a, b, c within this chain. The elements T_{ik} are usually simple trigonometric functions.

If we irradiate the sample with polarized radiation where the electric vector is parallel to the x-axis, only the component M_x of the individual transition moment will contribute to the absorption and its intensity will be proportional to M_x^2. The

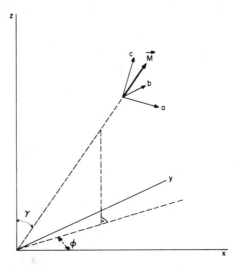

FIG. 7. Space and chain coordinate system: z-axis might be draw direction; c-axis might be identical with chain axis.

total absorption intensity A_x (optical density) is obtained by integrating M_x^2 over all the individual transition moments in an actual sample. In Section 4 we will evaluate (T) and carry out this integration separately for a number of different types of orientation distribution functions.

C. The Absorption Intensity Ellipsoid

At this point we will discuss this integration (described above) in general terms and show that with certain restrictions on the distribution function the absorption intensity for all polarization directions can be expressed by an equation describing an ellipsoid.

Let the distribution function $a(\gamma, \varphi)$ be the number of transition moments pointing in the segment γ, $\gamma+d\gamma$; φ, $\varphi+d\varphi$ (Fig. 7). The intensity A_x, for example, is then given by an integration of the form

$$A_x = \int\limits_{\gamma=0}^{\pi} \int\limits_{\varphi=0}^{2\pi} a(\gamma, \varphi) \sin \gamma M_x^2 \, d\varphi \, d\gamma \tag{14}$$

By such an integration it is not only possible to calculate A_x, A_y, and A_z but also $A_{\alpha,\beta,\chi}$, that is, the relative intensity (or extinction coefficient) for a band where the electric vector of the polarized radiation points in any direction, such as the

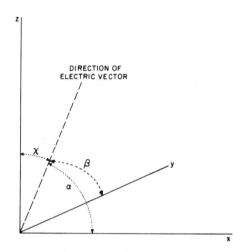

FIG. 8. Electric vector of polarized radiation in space coordinate system xyz.

α, β, χ direction as shown in Fig. 8. Since the component of an individual transition moment in this direction is given by

$$M_{\alpha,\beta,\chi} = M_x \cos \alpha + M_y \cos \beta + M_z \cos \chi \tag{15}$$

an integration similar to (14) leads to

$$A_{\alpha,\beta,\chi} = \int\limits_{\gamma=0}^{\pi} \int\limits_{\varphi=0}^{2\pi} a(\gamma, \varphi) \sin \gamma M_{\alpha,\beta,\chi}^2 \, d\varphi \, d\gamma \tag{16}$$

Combining Eqs. (14) and (15) with (16) we obtain

$$A_{\alpha,\beta,\chi} = A_x \cos^2 \alpha + A_y \cos^2 \beta + A_z \cos^2 \chi + 3 \text{ cross terms} \tag{17}$$

The cross terms in Eq. (17) are of the form

$$\int\limits_{\gamma=0}^{\pi} \int\limits_{\varphi=0}^{2\pi} a(\gamma, \varphi) \sin \gamma M_x M_y \cos \alpha \cos \beta \, d\varphi \, d\gamma \tag{18}$$

Cross terms of this type disappear if the distribution function for the transition moments $a(\gamma, \varphi)$ fulfills certain symmetry requirements which we now will discuss. From Fig. 7 it is obvious that M_x and M_y are given by

$$M_x = |\mathbf{M}| \sin \gamma \cos \varphi$$
$$M_y = |\mathbf{M}| \sin \gamma \sin \varphi$$

and the integrand in expression (18) becomes

$$a(\gamma, \varphi) \sin^3 \gamma \cos \varphi \sin \varphi |\mathbf{M}|^2 \cos \alpha \cos \beta$$

Let us consider the integration over φ from 0 to 2π. The function $\cos \varphi \sin \varphi$ is shown in Fig. 9. The integral over the product $a(\gamma, \varphi) \cos \varphi \sin \varphi$ will disappear if

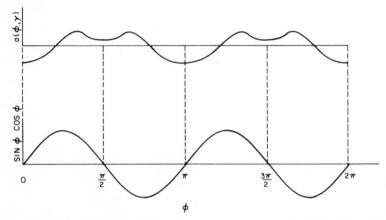

FIG. 9. General distribution function $a(\gamma, \varphi)$ which leads to an intensity ellipsoid compared with the function $\sin \varphi \cos \varphi$.

$a(\gamma, \varphi)$ is an even function in φ for $\varphi = 0$, $\pi/2$, π, and $3\pi/2$, in other words, if the xz- and yz-planes are planes of symmetry for the distribution function. An example for such a function is shown in Fig. 9.

Every distribution function, even the most asymmetric one, has an inversion center of symmetry, since in an absorption process, a transition moment is equivalent with its opposite one. Two planes and a center will create additional symmetry elements and lead to the point group D_{2h} (see e.g. Ref. 31) consisting of three planes, three twofold axes, the inversion center, and unity. For most practical cases the transition moment distribution function has this symmetry (or a higher one) and the cross terms in (17) disappear. Equation (17) is then the equation of an ellipsoid where $(A_x)^{-1/2}$, $(A_y)^{-1/2}$, and $(A_z)^{-1/2}$ are the lengths of the three main axes and $(A_{\alpha, \beta, \chi})^{-1/2}$ is the distance from the center O to a point P at the surface of the ellipsoid as shown in Fig. 10. In such a case OP is parallel to the electric vector of the incident radiation while the direction of propagation can be anywhere in a plane perpendicular to OP.

The existence of such an intensity ellipsoid means that the dichroic behavior of a sample can be defined by only three parameters such as the three intensities in the directions of the main axes of the ellipsoid. In other words, once the absorption intensities in the three main axes have been measured one does not gain any additional information with polarization measurements in various other directions (Ref. 57), but sometimes it might be more convenient to measure the intensity in a direction other than that of a main axis. Such measurements will be discussed in the following subsection.

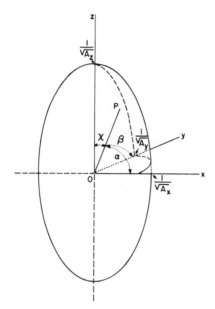

FIG. 10. Intensity ellipsoid.

D. DICHROIC RATIOS, GENERAL DEFINITIONS

In Section 2D we have defined the dichroic ratio for a certain band as the absorption intensity ratio for two perpendicular polarization directions. Now, dichroic ratios will be defined in more general terms by the following expressions:

$$R_{x,y} = \frac{1}{R_{y,x}} = \frac{A_x}{A_y} \tag{19a}$$

$$R_{y,z} = \frac{1}{R_{z,y}} = \frac{A_y}{A_z} \tag{19b}$$

$$R_{z,x} = \frac{1}{R_{x,z}} = \frac{A_z}{A_x} = \frac{R_{z,y}}{R_{x,y}} \tag{19c}$$

and for any two directions

$$R_{\alpha\beta\chi,\,\alpha'\beta'\chi'} = \frac{A_{\alpha\beta\chi}}{A_{\alpha'\beta'\chi'}} \tag{19d}$$

From Eqs. (19a–c) it is obvious that only two of the three dichroic ratios are mutually independent, which means that two dichroic ratios properly measured can fully describe the dichroic behavior of an absorption band.

As mentioned above, it is sometimes convenient to measure intensities in directions other than a main axis of the intensity ellipsoid. Two such cases will now be discussed, and we will show how the three dichroic ratios measured along the main axes can be computed from the observed data.

Dichroic Ratio at an Angle Different from 90°

Consider an oriented film in the plane of the paper which coincides with the xz-plane of the intensity ellipsoid. The radiation is propagated along the y-axis (perpendicular to the film) and polarized with the electric vector either in the

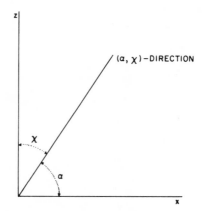

Fig. 11. Film in xz-plane, electric vector in z- and (α, χ)-direction.

z- or (α, χ)-direction as shown in Fig. 11 ($\beta = 90°$). In analogy to (19d) an observed dichroic ratio of the following form can then be defined:

$$R_{\alpha\chi,\,z} = \frac{A_{\alpha\chi}}{A_z} \tag{20}$$

Equations (17) and (20) lead to

$$R_{\alpha\chi,\,z} = \frac{A_x \cos^2\alpha + A_z \cos^2\chi}{A_z} = \frac{A_x}{A_z}\sin^2\chi + \cos^2\chi \tag{21}$$

Combining (21) with (19c) we obtain for the "right angle" dichroic ratio

$$R_{x,\,z} = \frac{R_{\alpha\chi,\,z} - \cos^2\chi}{\sin^2\chi} \tag{22}$$

Sample Tilted with Respect to the Radiation Beam

Let us consider a thin oriented film in the xy-plane. With radiation traveling perpendicular to this plane and polarized in various directions it is possible to determine the relative band intensities A_x and A_y or the respective dichroic ratio $R_{x,y}$. In order to characterize fully the intensity ellipsoid we have to measure the third component A_z or at least its relative magnitude, that is, a dichroic ratio where the component A_z is involved, such as $R_{z,y}$ [see Eq. (19)]. This could be done by cutting a thin section through the film, for example, parallel to the xz-plane, and analyzing it in the infrared microscope. Such a procedure is usually very tedious,

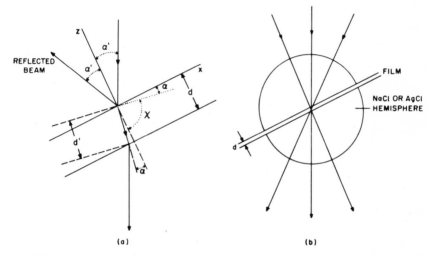

FIG. 12. (a) Film sample tilted in radiation beam. (b) Film between NaCl or AgCl hemispheres to reduce the amount of reflected radiation.

and it is easier to tilt the sample by the angle α' as shown in Fig. 12a. The optical densities A are measured for two polarization directions. For A_y the electric vector is parallel to the y-axis and for $A_{\alpha\chi}$ it is parallel to the dotted line (in Fig. 12a) which forms the angle α with the x-axis. α is defined by

$$\sin \alpha = \frac{\sin \alpha'}{n} \qquad (23)$$

where n is the refractive index of the sample. From Eq. (17) we obtain for $A_{\alpha\chi}$

$$A_{\alpha\chi} = A_x \cos^2 \alpha + A_z \sin^2 \alpha \qquad (24)$$

and with the definitions in (19) the observed dichroic ratio becomes

$$R_{\alpha\chi,y} = \frac{A_{\alpha\chi}}{A_y} = R_{x,y} \cos^2 \alpha + R_{z,y} \sin^2 \alpha \qquad (25)$$

$R_{x,y}$ can be measured by placing the film perpendicular to the light beam as described above. $R_{z,y}$ can be calculated from (25):

$$R_{z,y} = \frac{R_{\alpha\chi,y} - R_{x,y}\cos^2\alpha}{\sin^2\alpha} \tag{25a}$$

There are several difficulties involved with tilted sample measurements. One of them is the high loss due to reflection. This is particularly severe for the y-component of the polarized radiation while the (α,χ)-component is not reflected

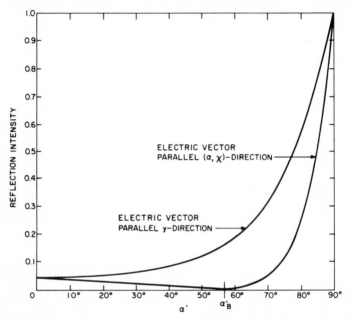

FIG. 13. Reflection intensity for two beams with perpendicular polarization directions, calculated according to Fresnel's equations (see, e.g., ref. 14) for a refractive index of 1.5.

at all if the sample is tilted at Brewster's angle. To reduce the reflected radiation, Tink (Ref. 56) has placed the sample between two hemispheres made of a transparent material such as sodium chloride or silver chloride as shown in Fig. 12b. All the rays of the slightly converging radiation beam hit the spherical surface at right angles where the reflection is a minimum and independent of the polarization direction. Furthermore, the reflection at the sample surface is much smaller than in Fig. 12a since the refractive index of the hemisphere material n_h is much closer to that of the sample than the refractive index of air which is assumed to be 1. In this case n in Eq. (23) has to be replaced by n/n_h.

Another difficulty is the determination of the refractive index of the sample, n, which will influence the angle α [Eq. (23)] and in turn the dichroic ratio $R_{z,y}$ as determined by Eq. (25a). The refractive index at a wavelength where the sample does not absorb (background for an absorption band) can be calculated from Eq. (1) where Brewster's angle α'_B has been obtained experimentally by determining the angle for a maximum intensity of the transmitted radiation polarized parallel to the (α, χ)-direction (Figs. 12 and 13). But at the absorption band itself the refractive index of the sample will change as a function of wavelength according to the dispersion theory (see e.g. Ref. 14). Furthermore, for a highly dichroic sample the refractive index is not the same for the two polarization directions. This will cause the effective absorption path lengths d' (Fig. 12a) to be different for the two polarization directions. In this way an error in the dichroic ratio is introduced since it is calculated as the ratio of two optical densities corresponding to two different path lengths. A correction for this error can be obtained only from a knowledge of the refractive index at the absorption peak which could be obtained from a dispersion formula or from reflection intensity measurements and Fresnel's equations (Ref. 14).

4. Dichroic Behavior of Individual Types of Orientation

A. General Discussion

High polymer samples can assume various types and degrees of orientation, ranging from a random distribution of polymer chains to a perfectly oriented sample comparable to a single crystal. Partial orientation can be observed in naturally occurring high polymers but the orientation effects are of particular interest in synthetic polymers in the form of fibers and films where the molecular chains have been oriented by mechanical deformations under suitable conditions. Such processes can, to a certain extent, be treated mathematically as shown by Kratky (Ref. 36) and Kuhn and Grün (Ref. 38). They have given a model describing the relation between stretching and orientation process and derived an expression for the expected distribution function for the chains. This model will be discussed in subsections E and F of this section.

In an unoriented sample the directional distribution of chains is random, which means that the number of chains $a(\gamma, \varphi) d\gamma d\varphi$ pointing into the space angle element φ, $\varphi + d\varphi$; γ, $\gamma + d\gamma$ is a constant. This is no longer true for oriented samples where more chains point in one direction than in another one. In such a case the spacial symmetry of $a(\gamma, \varphi)$ determines the *type* of orientation and the function itself determines the *amount* and perfection. It is useful to visualize the orientation type by its three-dimensional polar diagram, where the length of the vector pointing in the direction γ, φ is proportional to $a(\gamma, \varphi)$. This is shown in Table V where a classification of orientation types is given. Columns 2, 3, and 4 contain various possible shapes, description and symmetry elements respectively for the polar diagrams of each orientation type. For all types except No. 6, the absorption intensities for various polarization directions (direction of electric

TABLE   V

Classification of Orientation Types

Type of orientation	Examples of polardiagrams for chain orientation distribution	Characterization of polardiagram	Symmetry[a] axis of polardiagram		
			x	y	z
1. Random		Sphere	C_∞	C_∞	C_∞
2. Axial (a) perfect		All chains per-fectly aligned along z-axis	C_2	C_2	C_∞
(b) partial	 I II III IV	Chains preferentially aligned along z-axis, rotational sym-metry around z for Exs. I, II, & III	C_2	C_2	C_∞
		z is 4-fold symmetry axis for Ex. IV	C_2	C_2	C_4
3. Uniaxial (a) perfect	Same as 2a	Same as 2a, but chains are rotated around their own axis so that all equivalent functional groups or crystallite faces point in the same direction	Same as 2a		
(b) partial	Same as 2b	Same as 2b but ori-entation of mole-cules around chain axis as described in 3a. This orienta-tion may or may not be perfect	Same as 2b		

4. Planar
    (a) symmetrical
         perfect
	
	I II
		Ex. I. Circular disk in xy-plane. All polymer chains ran-domly distributed in this plane	C_2	C_2	C_∞

		Ex. II. Disk in xy-plane can have any shape as long as z is a 4-fold axis	C_2	C_2	C_4
(b) preferred perfect	 I II	All chains are parallel to xy-plane but preferentially aligned, e.g., in y-direction; z-axis is a 2-fold symmetry axis	C_2	C_2	C_2

[a] All orientation types have an inversion center. In types 1–5 the xy-, xz-, yz-planes are planes of symmetry.
[b] For definition see Eq. (26). M = Transition moment vector.

Main axis[b] and shape of intensity ellipsoid if transition moment with respect to chain axis is:			Equations in text describing:	
$\| \|$-	$\perp$	At an angle θ	Intensity ellipsoid	Dichroic ratios and other relations
$E_x = E_y = E_z$ Sphere	$E_x = E_y = E_z$ Sphere	$E_x = E_y = E_z$ Sphere		
$E_x = E_y = \infty$ E_z is finite Infinite disk	$E_x = E_y$ $E_z = \infty$ Infinite cylinder	$E_x = E_y \neq E_z$ Rotational ellipsoid	29 32	30 33
$E_x = E_y > E_z$ Flat rotational ellipsoid	$E_x = E_y < E_z$ Elongated rotational ellipsoid	$E_x = E_y \neq E_z$ Rotational ellipsoid	36-38 Orientation parameter for special types of orientation 40-51, 61	39 46
$E_x = E_y = \infty$, E_z is finite Infinite disk	M, e.g., in x-direction: $E_x, E_y = E_z = \infty$ Infinite disk in yz-plane	M, e.g., in xz-plane: $E_x, E_y = \infty, E_z$ Cylinder with elliptical cross section	88	
$E_x = E_y > E_z$ Flat rot. ellipsoid	M, e.g., preferentially in x-direction: $E_x < E_y, E_x < E_z$, usually $E_y \neq E_z$ Flat ellipsoid in yz-plane	$E_x \neq E_y \neq E_z$	73 where $f(\gamma, \varphi)$ is replaced by $f(\gamma, \varphi, \vartheta)$	
$E_x = E_y, E_z = \infty$ Infinite cylinder in z-direction	$E_x = E_y = \sqrt{2}E_z$ Flat rot. ellipsoid in xy-plane	$E_x = E_y \neq E_z$ Rot. ellipsoid around z-direction	63, 46	64 67-69
Same as Ex. I	Same as Ex. I	Same as Ex. I		
$E_x > E_y, E_z = \infty$ Infinite cylinder in z-direction with elliptical cross section	$E_y > E_x > E_z$	$E_x \neq E_y \neq E_z$	79	80, 81

TABLE V

Type of orientation	Examples of polardiagrams for chain orientation distribution	Characterization of polardiagram	Symmetry[a] axis of polardiagram		
			x	y	z
4. Planar *(Continued)* (c) symmetrical partial		Exs. I & II: Chains are partially aligned in xy-plane; rotational symmetry around z-axis	C_2	C_2	C_∞
		Ex. III. Same as Exs. I & II but z-axis is only a 4-fold symmetry axis	C_2	C_2	C_4
(d) preferred partial		Chains are partially aligned in xy-plane but preferentially in y-direction; x, y, & z are 2-fold symmetry axes	C_2	C_2	C_2
5. Uniplanar (a) symmetrical perfect	Same as 4a	Same as 4a*	Same as 4a		
(b) preferred perfect	Same as 4b	Same as 4b*	Same as 4b		
(c) symmetrical partial	Same as 4c	Same as 4c*	Same as 4c		
(d) preferred partial	Same as 4d	Same as 4d*	Same as 4d		
		* Individual chains or crystallites rotated around their own axis with specific functional group or crystallite face, respectively, pointing in preferred direction			
6. Unsymmetrical		No symmetry in distribution function except for center of inversion	None		

[a] All orientation types have an inversion center. In types 1–5 the xy-, xz-, yz-planes are planes of symmetry.
[b] For definition see Eq. (26). **M** = Transition moment vector.

(Continued)

Main axis[b] and shape of intensity ellipsoid if transition moment with respect to chain axis is:			Equations in text describing:	
$\|\|$	$\perp$	At an angle θ	Intensity ellipsoid	Dichroic ratios and other relations
$E_x = E_y < E_z$	$E_x = E_y > E_z > \dfrac{E_x}{\sqrt{2}}$	$E_x = E_y \neq E_z$	70, 38	71, 72
Rot. ellipsoid around z-axis	Flat rot. ellipsoid in xy-plane	Rot. ellipsoid around z-axis		
Same as Exs. I & II	Same as Exs. I & II	Same as Exs. I & II		
$E_y < E_x < E_z$	$E_y > E_x > E_z$	$E_x \neq E_y \neq E_z$	86	83, 87
			73	76
$E_x = E_y,\ E_z = \infty$	**M** in xy-plane: $E_x = E_y,\ E_z = \infty$; **M** $\|\|$ z-axis: $E_x = E_y = \infty,\ E_z$; **M** at angle $\neq 90°$ with z-axis: $E_x = E_y \neq E_z$	$E_x = E_y \neq E_z$		
$E_x > E_y,\ E_z = \infty$	**M** in xy-plane: $E_x < E_y,\ E_z = \infty$; **M** $\|\|$ z-axis: $E_x = E_y = \infty,\ E_z$; **M** at angle $\neq 90°$ with z-axis: $E_x < E_y \neq E_z$	$E_x \neq E_y \neq E_z$		
$E_x = E_y < E_z$	**M** preferentially in xy-plane: $E_x = E_y < E_z$; **M** preferentially $\|\|$ z-axis: $E_x = E_y > E_z$; **M** preferentially at angle $\neq 90°$ with z-axis: $E_x = E_y \neq E_z$	$E_x = E_y \neq E_z$		
$E_y < E_x < E_z$	**M** preferentially in xy-plane: $E_x < E_y < E_z$; **M** preferentially $\|\|$ z-axis: $E_y > E_x > E_z$; **M** preferentially at angle $\neq 90°$ with z-axis: $E_x \neq E_y \neq E_z$	$E_x \neq E_y \neq E_z$		

No intensity ellipsoid

vector) are characterized by intensity ellipsoids (see Section 3C) which are determined by the three main axes E_x, E_y, and E_z where

$$E_x = \frac{1}{\sqrt{A_x}}; \qquad E_y = \frac{1}{\sqrt{A_y}}; \qquad E_z = \frac{1}{\sqrt{A_z}} \qquad (26)$$

The relative magnitudes of E_x, E_y, and E_z are given in columns 5, 6, and 7 of Table V.

As shown in Sections 3B and 3C the expected absorption intensity is obtained by an integration procedure. It can be seen from Table V that quite a number of different orientation types can lead to the same intensity ellipsoid. For example, a sample with perfect axial orientation (2a) and a transition moment at an unknown angle $\theta \neq 90°$ to the chain axis might have exactly the same intensity ellipsoid as a sample with partial symmetrical planar orientation (4c). Infrared dichroic measurements on one absorption band alone could not distinguish between the two; additional information is, therefore, required. Absorption bands with the transition moment parallel (or perpendicular) to the chain axis would have different intensity ellipsoids for the orientation types shown in 2a and 4c. In many cases the direction of the transition moment relative to the chain axis is approximately known, and it is possible to decide between various orientation types by measuring the dichroism of a number of bands. The geometrical shape and history of a sample can also help to decide between various types of orientation. A drawn fiber, for example, usually shows axial orientation while a two-way stretched film is likely to be planar oriented.

It is important to emphasize again that the observed dichroism is obtained by *integration* over all actual chain orientations and it usually cannot give any information about the distribution function itself except in an indirect way, that is, one can assume a certain distribution function, carry out the integration and compare the results with actual measurements. But there are many other intensity distributions leading to the same intensity ellipsoid. A perfect symmetrical planar distribution as shown in 4a I has the same dichroic behavior as 4a II. In particular for light traveling along the z-axis the intensity is the same for any polarization direction even though the distribution of chains in the xy-plane is not homogeneous for 4a II. An X-ray or electron diffraction diagram on the other hand would show a difference between I and II since the intensity distribution along a diffraction arc is a measure of the number of chains pointing in a given direction.

In the following paragraphs we will discuss individual examples of certain types of orientation and calculate the expected intensity ellipsoid and dichroic ratios. In some specific cases it is possible to compute the angle θ of the transition moment with the polymer chain axis. Most of the common types of orientation will be described in detail. The general methods will, of course, also be applicable to very specific types not treated in this book (see e.g. Ref. 35).

B. Perfect Axial Orientation

Transition Moment Fixed with Respect to the Chain Axis

In this type of orientation (2a in Table V) the molecular chains are all aligned parallel, but they are distributed at random with respect to a rotation around their own axis c. A highly drawn and oriented polymer filament with a round cross section usually shows this type of orientation, but it also might be present in other samples such as film strips highly drawn in one direction.

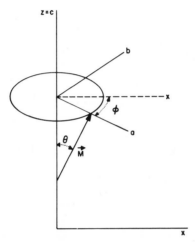

Fig. 14. Perfect axial orientation, chain axis is parallel to z-axis. Transition moment **M** forms angle θ with chain axis.

The relative position of the chain (a, b, c) and space (x, y, z) coordinate systems is shown in Fig. 14 for an individual chain with a transition moment **M** forming an angle θ with the chain axis c. The components of **M** in the chain coordinate system are

$$M_a = M \sin \theta, \qquad M_b = 0, \qquad M_c = M \cos \theta \qquad (27)$$

where $M = |\mathbf{M}|$, and the components in the space coordinate system are given by the transformation equation (13) which in this specific case assumes the following form:

$$(M_x, M_y, M_z) = \begin{pmatrix} \cos \varphi & \sin \varphi & 0 \\ -\sin \varphi & \cos \varphi & 0 \\ 0 & 0 & 1 \end{pmatrix} \begin{pmatrix} M_a \\ M_b \\ M_c \end{pmatrix} \qquad (28)$$

The absorption intensity for an individual group in any one direction, for example, the x-direction, is proportional to the square of the component of the transition moment M_x given in (28). The expected band intensity A_x for the whole sample is obtained by integrating M_x^2 over all possible chain orientations. For the specific case of perfect axial orientation the integration is very simple since it has to be carried out over the angle φ only. From (27) and (28) we obtain

$$A_x = \frac{1}{2\pi} \int_0^{2\pi} (\cos \varphi M_a + \sin \varphi M_b)^2 \, d\varphi$$

$$= \frac{1}{2\pi} M^2 \sin^2 \theta \int_0^{2\pi} \cos^2 \varphi \, d\varphi = \tfrac{1}{2} M^2 \sin^2 \theta \qquad (29)$$

$$A_y = A_x \qquad \text{(for symmetry reasons, see Table V)}$$

$$A_z = \frac{1}{2\pi} \int_0^{2\pi} M_c^2 \, d\varphi = M^2 \cos^2 \theta$$

$1/2\pi$ is a normalization factor.

The expected dichroic ratios as defined by (19) are given by the following expressions (Ref. 26)

$$R_{x,y} = 1; \qquad R_{x,z} = R_{y,z} = \tfrac{1}{2}\tan^2 \theta \qquad (30)$$

For a transition moment that is parallel to the chain axis ($\theta = 0$) $R_{x,z}$ becomes 0 and for a perpendicular band ($\theta = 90°$) $R_{x,z}$ is equal to infinity. It is interesting to note that $R_{x,z} = 1$ if $\theta = 54° 44'$. No dichroism can be observed for such an absorption band since its intensity in any polarization direction is the same.

If one knows that a given sample has perfect axial orientation the angle θ for various transition moments can be calculated from the observed dichroic ratios of the corresponding absorption bands. This is particularly useful for cases where the transition moment is intimately connected with a chemical group such as $>CH_2$, $—C\equiv N$, etc., and it is possible to obtain information about the geometrical structure of the polymer chain.

Transition Moment with a Rotational Degree of Freedom

As an example we consider polystyrene where the benzene ring is one C—C bond length removed from the backbone chain as shown in Fig. 15a. We assume that this bond $C_1—C_2$ forms an angle θ with the chain axis and that the benzene ring can rotate at random around this bond. Under these conditions we will calculate the expected dichroism for an absorption band caused by a benzene ring transition moment **M** forming the angle Ψ with the $C_1—C_2$ bond.

In Fig. 15b the sample coordinate system has been chosen similar to Fig. 14 so that the transformation equation (28) also holds for this case. The components of **M** in the chain system are

$$M_a = M(\cos \Psi \sin \theta - \sin \Psi \cos \vartheta \cos \theta)$$
$$M_b = -M \sin \Psi \sin \vartheta \qquad (31)$$
$$M_c = M(\cos \Psi \cos \theta + \sin \Psi \cos \vartheta \sin \theta)$$

An integration similar to (29) will now have to be carried out over the angles φ as well as ϑ from 0 to 2π. In analogy to (29) we obtain

$$A_y = A_x = \frac{1}{4\pi^2} M^2 \int_0^{2\pi} \int_0^{2\pi} [\cos\varphi \, (\cos\Psi \sin\theta - \sin\Psi \cos\vartheta \cos\theta)$$

$$- \sin\varphi \sin\Psi \sin\vartheta]^2 \, d\varphi \, d\vartheta$$

$$= \tfrac{1}{4} M^2 (2\cos^2\Psi \sin^2\theta + \sin^2\Psi \cos^2\theta + \sin^2\Psi) \qquad (32)$$

$$A_z = \frac{1}{4\pi^2} M^2 \int_0^{2\pi} \int_0^{2\pi} (\cos\Psi \cos\theta + \sin\Psi \cos\vartheta \sin\theta)^2 \, d\varphi \, d\vartheta$$

$$= \tfrac{1}{2} M^2 (2\cos^2\Psi \cos^2\theta + \sin^2\Psi \sin^2\theta)$$

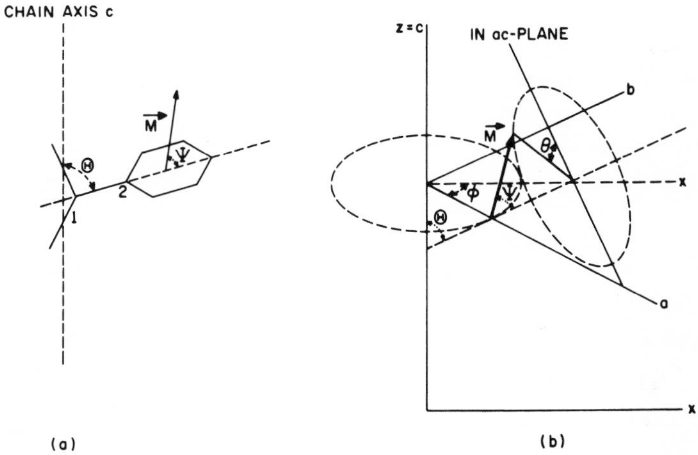

FIG. 15. Transition moment with two degrees of freedom relative to the chain axis. (a) Segment of a polystyrene chain, transition moment **M** at an angle Ψ with respect to the plane of the benzene ring. (b) General model.

These equations lead to the following dichroic ratios:

$$R_{x,y} = 1$$

$$R_{x,z} = R_{y,z} = \frac{2\cos^2\Psi \sin^2\theta + \sin^2\Psi \cos^2\theta + \sin^2\Psi}{4\cos^2\Psi \cos^2\theta + 2\sin^2\Psi \sin^2\theta} \qquad (33)$$

Equation (33) is more complex than (30) since the expected dichroic ratio depends on two variables. This means that it is no longer possible to deduce the geometrical location of the transition moment from the value of $R_{y,z}$ unless one of the two angles Ψ and θ is known from an additional source of information.

It was pointed out by L. F. Beste (Ref. 8) that $R_{x,z} = 1$ if either θ or Ψ is

54°44′ [see comment following Eq. (30)]. An equation for $R_{x,z}$ can also be derived if an additional rotational degree of freedom, characterized by an angle Φ is introduced. For this case too it can be shown (Ref. 8) that $R_{x,z} = 1$ if either θ, Ψ, or Φ is 54° 44′.

Equation (33) goes over into (30) if $\Psi = 0$. Such a behavior is expected since Fig. 15b is equivalent to Fig. 14 under this condition. In many practical cases a group removed from the chain by one bond length cannot rotate freely around this bond, and, therefore, will assume certain preferred locations. In such a case (33) would no longer be valid.

C. PARTIAL AXIAL ORIENTATION

Expected Intensity Ellipsoid and Dichroic Ratio

This type of orientation is listed under 2b in Table V. The chains are preferentially aligned along the z-axis which is at least a four-fold symmetry axis, and in most cases the distribution of chains is rotational symmetrical about z (Fig. 16).

As in the case of perfect axial orientation we consider a chain with an axis c and a transition moment **M** at an angle θ with respect to c. The molecular coordinate system is again chosen so that **M** is in the ac-plane. The axis c itself is no longer parallel to z but at an angle γ which can vary between 0 and $\pi/2$, and the number of chains pointing in a certain direction is given by a distribution function $f(\gamma)$.

In Section 3C a slightly different distribution function $a(\gamma, \varphi)$ was defined. For clarification we briefly compare the two since both of them are used in the literature. $a(\gamma, \varphi)\,d\gamma\,d\varphi$ is proportional to the number of chains pointing into the shaded angle element in Fig. 17, while for a rotational symmetrical distribution $f(\gamma)\,d\gamma$ is a partially integrated distribution function. It is proportional to the number of chains pointing into the dotted angle element. The two functions are related by

$$f(\gamma) = \frac{1}{2\pi} \int\limits_{0}^{2\pi} a(\gamma, \varphi) \sin \gamma \, d\varphi = a(\gamma, \varphi) \sin \gamma \tag{34}$$

This relation only holds for a rotational symmetrical distribution where $a(\gamma, \varphi)$ is independent of φ. Figure 18 shows some examples. The dashed lines are the distribution functions for an unoriented sample. Curves I and II are $f(\gamma)$ functions for samples having medium and high axial orientations, respectively.

We now will calculate the main axis for the expected intensity ellipsoid. Equation (27) also holds for this case because of the same choice of the coordinate system (a, b, c) as in Fig. 14, but the transformation equation (13) now assumes a more complex form:

(M_x, M_y, M_z)

$$= \begin{pmatrix} \cos\vartheta\cos\gamma\cos\varphi + \sin\vartheta\sin\varphi & \sin\vartheta\cos\gamma\cos\varphi - \cos\vartheta\sin\varphi & \cos\varphi\sin\gamma \\ \cos\vartheta\cos\gamma\sin\varphi - \sin\vartheta\cos\varphi & \sin\vartheta\cos\gamma\sin\varphi + \cos\vartheta\cos\varphi & \sin\varphi\sin\gamma \\ -\sin\gamma\cos\vartheta & -\sin\gamma\sin\vartheta & \cos\gamma \end{pmatrix} \begin{pmatrix} M_a \\ M_b \\ M_c \end{pmatrix} \tag{35}$$

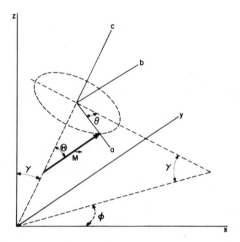

FIG. 16. Partial axial orientation, chain axis c forms angle γ with z-axis.

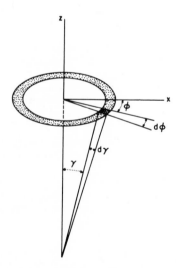

FIG. 17. Comparison of the distribution function $a(\gamma, \varphi)$ with $f(\gamma)$ which is partially integrated (see text).

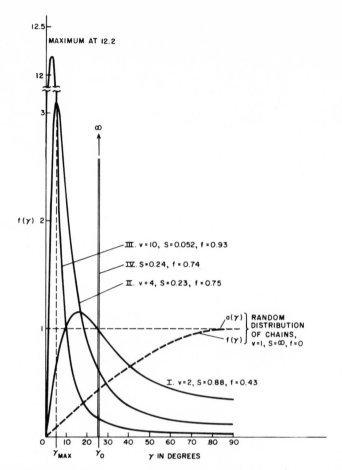

FIG. 18. Distribution functions $f(\gamma)$ for axial orientation. I, II, and III are calculated according to Kratky's model [Eq. (57)]. v is the draw ratio; S and f are orientation parameters (see text).

The expected absorption intensities A for a band corresponding to the transition moment **M** are given by the integrations†

$$A_y = A_x \frac{1}{4\pi^2} \int_{\gamma=0}^{\pi/2} \int_{\varphi=0}^{2\pi} \int_{\vartheta=0}^{2\pi} M_x^2(\gamma, \vartheta, \varphi) f(\gamma) \, d\vartheta \, d\varphi \, d\gamma \qquad (36)$$

$$A_z = \text{same integral over } M_z^2$$

† In (14) the integration over γ was carried out from 0 to π. Here we integrate only from 0 to $\pi/2$ since the integrand is always a symmetric function with respect to a plane perpendicular to the z-axis.

From (35) and (36) we obtain

$$A_y = A_x = \frac{M^2}{4\pi^2} \int_0^{\pi/2} \int_0^{2\pi} \int_0^{2\pi} [(\cos \vartheta \cos \gamma \cos \varphi + \sin \vartheta \sin \varphi) \sin \theta \\ + \cos \varphi \sin \gamma \cos \theta]^2 f(\gamma) \, d\vartheta \, d\varphi \, d\gamma \quad (37)$$

$$A_z = \frac{M^2}{4\pi^2} \int_0^{\pi/2} \int_0^{2\pi} \int_0^{2\pi} [-\sin \gamma \cos \varphi \sin \theta + \cos \gamma \cos \theta]^2 f(\gamma) \, d\vartheta \, d\varphi \, d\gamma$$

The integration can be carried out over the angles ϑ and φ which leads to the following expressions:

$$A_y = A_x = \frac{M^2}{4} \left[\sin^2 \theta \int_0^{\pi/2} (2 - 3 \sin^2 \gamma) f(\gamma) \, d\gamma + \int_0^{\pi/2} 2 \sin^2 \gamma f(\gamma) \, d\gamma \right]$$

$$\quad (38)$$

$$A_z = \frac{M^2}{2} \left[\cos^2 \theta \int_0^{\pi/2} (2 - 3 \sin^2 \gamma) f(\gamma) \, d\gamma + \int_0^{\pi/2} \sin^2 \gamma f(\gamma) \, d\gamma \right]$$

The dichroic ratio $R_{x,z}$ can now easily be obtained from (38). With Beer (Ref. 6) we write it in the following form:

$$R_{x,z} = \frac{A_x}{A_z} = \frac{\sin^2 \theta + S}{2 \cos^2 \theta + S} \quad (39)$$

S is defined by

$$S = \frac{F}{N - \frac{3}{2}F}$$

where

$$\left. \begin{array}{l} \\ F = \int_0^{\pi/2} \sin^2 \gamma f(\gamma) \, d\gamma; \qquad N = \int_0^{\pi/2} f(\gamma) \, d\gamma \end{array} \right\} \quad (40)$$

The distribution function $f(\gamma)$ is usually normalized so that $N = 1$ by definition, and from (40) we obtain

$$S = \frac{F}{1 - \frac{3}{2}F} \quad (41)$$

Discussion of the Dichroic Ratio Formula

Let us now discuss Eq. (39) by pointing out the following facts.

1. The dichroic ratio $R_{x,z}$ is a simple function of the angle θ and an orientation parameter S (see Fig. 19) which only depends on the shape of the distribution function $f(\gamma)$. S varies from 0 to ∞ by going from a perfectly oriented sample to a random chain distribution and from $-\infty$ to -2 by going from the random distribution to a symmetrical perfect planar orientation (Fig. 21). If the distribution function $f(\gamma)$ is known, it is possible to calculate the parameter S by Eq. (40).

This can sometimes be done explicitly but in other cases only a numerical computation is possible. A number of individual distribution functions will be discussed below.

2. Equation (39) can be used in two different ways once the dichroic ratio $R_{x,z}$ has been determined experimentally. If S is known the angle θ can easily be calculated. Thus, information about the geometrical structure of the polymer

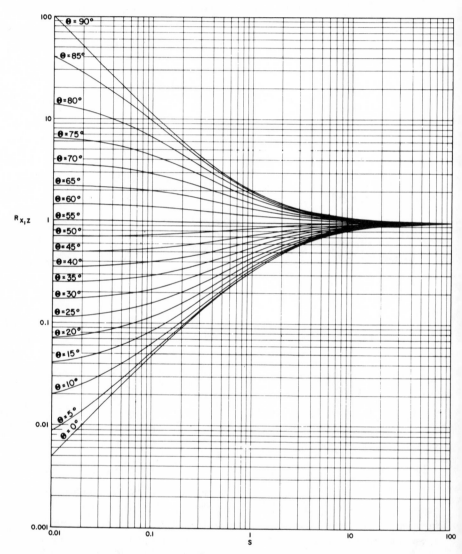

FIG. 19. Dichroic ratio as a function of the angle θ and the orientation parameter S according to Eq. (39).

chain can be obtained. If on the other hand the angle θ is known, it is possible to compute the orientation parameter S, which is an indirect measure for the perfection of the axial orientation. To facilitate such computations $R_{x,z}$ is plotted in Fig. 19 as a function of S and θ. The curves are given only for positive S values corresponding to a true axial orientation.

3. In many practical cases neither S nor θ are known and we have the problem of determining both. This is sometimes possible, at least to some extent, if several absorption bands at various frequencies can be measured. For a given homogenous sample the orientation parameter S is the same for all these bands. If, for example, one angle θ for one transition moment is known, S can be determined and since it has the same value for all the other bands θ can be calculated for those. In some cases it is possible to give an upper and lower limit for certain θ values by a method described by Fraser (Ref. 29). It is best to describe this method with the help of an example. Consider three absorption bands in a spectrum with $R_{x,z}$ values of $R_1 = 4$, $R_2 = 3$, $R_3 = 1.5$. Limits for the three values of θ can be given in the following way. Start out with the largest dichroic ratio R. From Fig. 19 or Eq. (39) we find that for $R_1 = 4$, the angle

$$\theta_1 > 71°$$

and the orientation parameter $S < 0.33$. This upper limit for S imposes an upper limit on the angles θ_2 and θ_3. For $R_2 = 3$ we obtain a lower limit of 63° for θ_2 and since S has to be smaller than 0.33 the upper limit is 78° (Fig. 19):

$$63° < \theta_2 < 78°$$

In analogy to this we obtain for θ_3:

$$60° < \theta_3 < 63°$$

This method only works for a homogenous sample where S is the same for all absorption bands.

4. Equation (39) has been derived for a distribution function which has rotational symmetry around the z-axis but it holds for any distribution function as long as it leads to a rotational symmetrical intensity ellipsoid. For example, a distribution function where z is a fourfold symmetry axis would fulfill this condition. This particular case will be of some importance in Sections 4D and 4E where symmetrical planar orientation will be discussed.

Various Degrees and Types of Axial Orientation

In this paragraph we will discuss a number of distribution functions $f(\gamma)$ which can be integrated so that the orientation parameter S can be calculated with the help of Eq. (40). Most of these functions have been discussed by Fraser (Ref. 28).

1. *Perfect axial orientation.* All chains are parallel to the z-axis and the distribution function is

$$f(\gamma) = \delta(\gamma) \tag{42}$$

where $\delta(x)$ is defined by

$$\delta(0) = \infty, \qquad \delta(x \neq 0) = 0,$$

$$\int\limits_{-\epsilon}^{+\epsilon} \delta(x)\,dx = 1 \tag{43}$$

where ϵ is a small number. From (40) we obtain $F = 0$ which leads to $S = 0$. Under these circumstances (39) is equivalent to (30) as expected for perfect axial orientation.

2. *Randomly oriented sample.* For a random distribution of chains the distribution function becomes

$$f(\gamma) = \sin\gamma \tag{44}$$

as shown in Fig. 18. Eq. (40) leads to $F = \frac{2}{3}$ and $S = \infty$. The dichroic ratio $R_{x,z} = 1$ according to Eq. (39). This behavior is expected for an unoriented sample.

3. *Perfect symmetrical planar orientation* (4a in Table V). This type will be discussed in the next section. It is only mentioned here as a limiting case for an axial distribution function.

$$f(\gamma) = \delta(90° - \gamma) \tag{45}$$

where $\delta(x)$ is defined by (43). With (40) we obtain $F = 1$, $S = 2$, and (39) becomes

$$R_{x,z} = \frac{\sin^2\theta - 2}{2\cos^2\theta - 2} \tag{46}$$

This equation is equivalent to (64) and will be discussed further in the section on planar orientation.

4. *Sample consisting of a random and a fully oriented fraction.* Fraser (Refs. 27, 28) and Beer (Ref. 7) have considered a polymer sample consisting of a randomly oriented portion r and a fully oriented (perfect axial) fraction $f = 1 - r$. The distribution function is, therefore, a mixture of (42) and (44) and has the form

$$f(\gamma) = r\sin\gamma + (r-1)\delta(\gamma) \tag{47}$$

Equations (40) and (47) lead to

$$N = 1; \qquad F = \frac{2}{3}r; \qquad S = \frac{2}{3}\frac{r}{1-r} = \frac{2}{3}\frac{1-f}{f} \tag{48}$$

Since r can vary from 0 to 1, S will vary between 0 and ∞.

A sample consisting of a fully oriented and a random fraction does not exist in reality, but it is a very useful model to help visualize the amount of orientation present in a sample. As an example, let us assume an orientation function $f(\gamma)$ as shown in curve I of Fig. 18. If this function is well known, Eq. (40) can be integrated and an S-value can be computed. For curve I the S-value turns out to be 0.88. On the other hand, we can ask which mixture of a randomly and fully oriented sample leads to the same S-value and, therefore, to the same dichroic ratio $R_{x,z}$. The answer is obtained by (48) from which r is calculated to be about 0.57. A sample consisting of 57% unoriented and 43% fully oriented chains would correspond to the sample with a distribution function $f(\gamma)$ as far as the dichroic behavior is concerned.

For many practical cases it is convenient to use r or f as orientation parameters, especially if the shape of the distribution function is not known. It still is possible to calculate S and subsequently r or f by Eqs. (39) and (48), respectively, for known values $R_{x,z}$ and θ. This enables one to express the amount of orientation

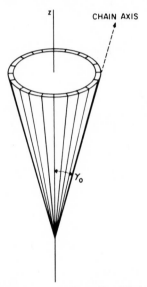

CHAIN AXIS

z

γ_0

FIG. 20. Distribution function given by Eq. (49). All the chains at an angle γ_0 with respect to the z-axis.

in a quantitative fashion. The number $100\,f$ (where $f = 1 - r$) can be called the "percent orientation number." We want to emphasize that this is an arbitrary definition and point out that various shapes of $f(\gamma)$ might lead to the same value of r or f.

In a previous subsection we have mentioned that for a partial symmetrical planar orientation (4c in Table V) S assumes a value between $-\infty$ and -2 which formally leads to r-values ranging from 1 to $\tfrac{3}{2}$. Such a value of r is meaningless since a fraction cannot be larger than one. But on a formal basis it still could be used to characterize the degree of planar orientation.

5. *All chains are oriented at a fixed angle.* The chain distribution of this type is shown in Fig. 20 and the distribution function $f(\gamma)$ is given by (curve IV in Fig. 18):

$$f(\gamma) = \delta(\gamma - \gamma_0) \tag{49}$$

Equations (49), (43), and (40) lead to (Ref. 7)

$$N = 1; \qquad F = \sin^2 \gamma_0 \tag{50}$$

$$S = \frac{2\sin^2 \gamma_0}{2 - 3\sin^2 \gamma_0} \tag{51}$$

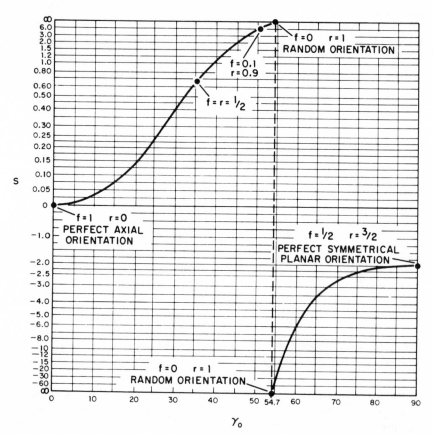

Fig. 21. Orientation parameter S as a function of the orientation angle γ_0. f and r values [Eq. (48)] are shown for a few points. (Note the difference in scale for positive and negative values of S.)

This type of chain distribution is again not realistic but it can be close to a real distribution especially if $f(\gamma)$ is a very sharp curve similar to II and III in Fig. 18 where in an approximation these curves can be replaced by the delta function (49) where $\gamma_0 = \gamma_{max}$.

For highly crystalline samples an average orientation angle γ_{av} can sometimes be determined by X-ray or electron diffraction methods. In such cases γ_0 can be chosen to be equal to γ_{av} and S-values can be calculated according to (51) which in turn can be used to compute θ by Eq. (39).

The S-values are plotted in Fig. 21 as a function of γ_0 according to Eq. (51). Some values of r and f are also marked on the same curve.

6. *Orientation obtained in a drawing process—Kratky's model.* Kratky (Ref. 36) and Kuhn and Grün (Ref. 38) have discussed a model for the orientation mechanism in a drawing process. Such a mechanism leads to a distribution function which is probably the most realistic one of those discussed so far. We, therefore, will briefly describe this model and calculate the orientation parameter S as a function of draw ratio.

We start out with an unoriented sample where the chain segments are distributed

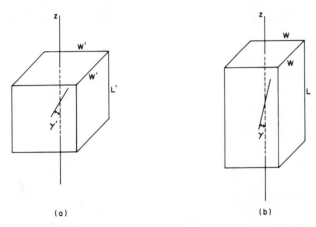

FIG. 22. Kratky's model (Ref. 36) for the orientation mechanism in a drawing process.

at random. The orientation mechanism can best be described with the help of Fig. 22. Figure 22a represents a volume element of the undrawn sample and a chain segment at an arbitrarily chosen angle γ' with the z-axis. Figure 22b shows the same volume element but drawn by the draw ratio v which is defined by

$$v = \frac{L}{L'} \tag{52}$$

If the density does not change, the volume of the elements in Figs. 22a and 22b is the same:

$$W^2 L = W'^2 L' \tag{53}$$

Furthermore it is assumed that the molecular orientation of a chain segment corresponds to the macroscopic deformation of the sample so that (52) and (53) lead to

$$\tan \gamma = \tan \gamma' \, v^{-3/2} \tag{54}$$

In Fig. 22a the number of chain segments $f(\gamma')$ in the interval γ' and $\gamma' + d\gamma'$ is equal to $a(\gamma')\sin\gamma'$ [see (34)]. The same chain segments point into the interval $\gamma, \gamma + d\gamma$ in Fig. 22b, so that

$$a(\gamma')\sin\gamma'\,d\gamma' = a(\gamma)\sin\gamma\,d\gamma \tag{55}$$

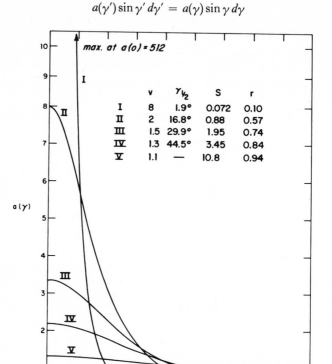

FIG. 23. Kratky distribution function $a(\gamma)$ [Eq. (57)] for various draw ratios v.

where $a(\gamma') = 1$ by definition (for a random distribution). From (54) we obtain

$$\frac{d\gamma}{d\gamma'} = \frac{(1+\tan^2\gamma')\,v^{-3/2}}{1+\tan^2\gamma'\,v^{-3}} \tag{56}$$

Combining (34), (54), (55), and (56) the distribution function becomes (Refs. 36, 38)

$$f(\gamma) = a(\gamma)\sin\gamma = \frac{v^{3/4}\sin\gamma}{(v^{-3/2}\cos^2\gamma + v^{3/2}\sin^2\gamma)^{3/2}} \tag{57}$$

Figure 18 shows this function for a number of draw ratios v. The distribution function $a(\gamma)$ is shown in Fig. 23. It corresponds almost exactly to the intensity distribution function in a meridianal arc of an X-ray or electron diffraction pattern if the orientation of crystallites only is considered. Such a distribution function is often characterized by its half width $\gamma_{1/2}$, that is the angle at which the distribution function assumes half its peak value.

$$a(\gamma_{1/2}) = \tfrac{1}{2}a(0) \tag{58}$$

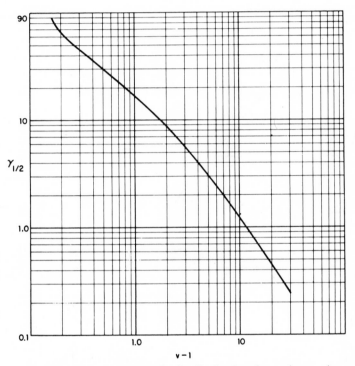

FIG. 24a. Half width of Kratky distribution function against $v-1$.

From (57) we easily obtain

$$a(0) = v^3 \tag{59}$$

Equations (57), (58), and (59) lead to

$$\sin^2 \gamma_{1/2} = \frac{\sqrt[3]{4}-1}{v^3-1} \tag{60}$$

Figure 24a shows the half width $\gamma_{1/2}$ as a function of the draw ratio v (for convenience the abscissa is $v-1$ instead of v).

The Kratky distribution function can be related to the orientation parameters S and r (or f) discussed in previous paragraphs, by carrying out the integrations of (40) with $f(\gamma)$ of (57). The results are (Ref. 29):

$$N = 1$$

$$F = 1 - \frac{v^3}{v^3 - 1} + \frac{v^3}{(v^3 - 1)^{3/2}} \cos^{-1} \frac{1}{v^{3/2}} \tag{61}$$

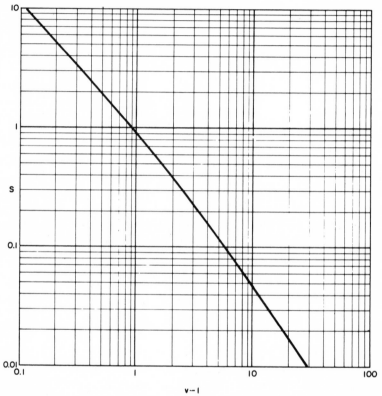

Fig. 24b. Orientation parameter S as a function of $v-1$ according to Kratky's theory.

The orientation parameter S can easily be calculated from (41) and (61). The result is plotted in Fig. 24b which can be used directly in practical calculations. For convenience the fractions r and f [see (48)] as a function of the draw ratio v are shown in Fig. 24c (Ref. 28) and $\gamma_{1/2}$ as a function of S in Fig. 24d. The orientation parameters are also given for various distribution functions plotted in Figs. 18 and 23.

A few critical remarks are necessary at this point. A "drawn polymer model" like the one just described leads to the most realistic distribution functions $f(\gamma)$ from which it is possible to calculate absolute values of the orientation parameters

S as a function of the draw ratio *v*. But even such a model is only an approximation to the real orientation mechanism which depends upon the polymer itself, as well as the drawing conditions. It is well known that samples drawn by the same amount (same *v*) can have different degrees of orientation. Deviations from the model are particularly expected if the drawing is carried out close to the polymer melting or softening temperature where viscous flow occurs, which in the extreme

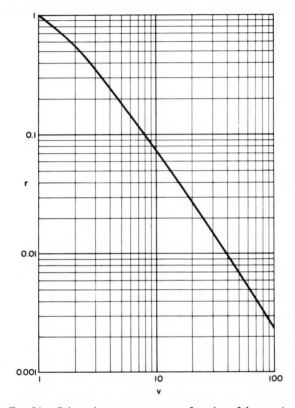

FIG. 24c. Orientation parameter *r* as a function of draw ratio *v*.

case does not result in any orientation at all. This means that the orientation calculated by (57) is the highest possible, so that the actual degree of orientation is always somewhat lower and the *S*-value obtained from Fig. 24b (for a given *v*) is a lower limit. The orientation can probably still be described by a function of the form of (57) even though some viscous flow occurred, but the value of *v* can then be considered only a parameter and not a "true" draw ratio (Ref. 29). Since the Kratky distribution function seems to be the most realistic one, it should be used if various methods for orientation measurement such as infrared, X-rays, electron diffraction, birefringence, or sonic velocity measurements are compared with each other.

7. *Other types of axial orientations.* A number of other types of distribution functions have been discussed in the literature mostly because they are simple mathematical functions and can easily be integrated. Keller and Sandeman (Ref. 35) have suggested a function $a(\gamma) = $ constant for $\gamma < \gamma_0$ and $a(\gamma) = 0$ for $\gamma > \gamma_0$, while Tobin and Carrano (Ref. 57) have discussed an ellipsoidal function for $a(\gamma)$.

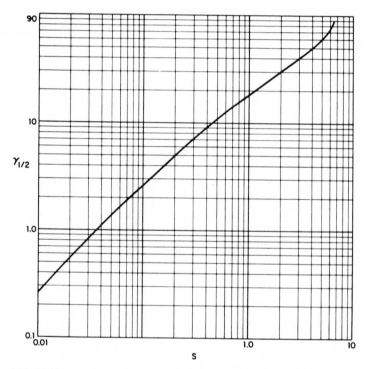

FIG. 24d. Half width of Kratky distribution function against orientation parameter S.

In these distributions γ_0 and the eccentricity of the ellipsoid, respectively, could be used as orientation parameters. They would not yield any more information than the parameters discussed previously, although the shape of an ellipsoid, for example, might help to visualize the degree of orientation.

Practical Example

Bohn (Ref. 13) has measured the dichroic behavior of polyacrylonitrile fibers as a function of draw ratio v. As an example we will discuss the dichroic ratio of the $-C{\equiv}N$ stretching vibrational band in the 2230 cm^{-1} region. We will show how the Kratky orientation model can be applied to this case of axial orientation in order to calculate the angle θ of the transition moment with the chain axis. The data are listed in Table VI, where the observed dichroic ratio values were corrected for spectral dilution, beam convergence, polarizer deficiency and insufficient resolution of the monochromator as discussed in Section 2D. The values of v have also been corrected for an initial orientation (due to spinning) in the undrawn fibers, corresponding to a draw ratio of 1.14. Also listed in Table VI are the orientation parameters S and r. They were obtained from Figs. 24b and 24c, respectively. The last column gives the value for the angle θ obtained from Fig. 19. It turns out that θ is approximately constant for all draw ratios. This is expected since θ should depend only on the geometrical structure of the chain itself but not upon its orientation.

TABLE VI

ORIENTATION MEASUREMENTS ON THE —C≡N STRETCHING
VIBRATIONAL BAND FOR DRAWN POLYACRYLONITRILE FIBERS

Draw ratio v	R	S	r	θ (deg)
2.28	1.93	0.660	0.49	73
4.56	2.40	0.182	0.21	68
5.70	2.68	0.125	0.16	69
6.85	2.91	0.093	0.12	69
7.98	2.96	0.073	0.10	69
9.13	3.24	0.060	0.08	70
11.40	3.51	0.042	0.06	71
13.70	3.82	0.032	0.05	71
18.28	4.12	0.020	0.03	71

D. PERFECT SYMMETRICAL PLANAR ORIENTATION (4a in Table V)

Planar orientation is very common in film samples which have been stretched in two directions. The type of orientation discussed in this and the following section is obtained if the two draw directions form a right angle and if the draw ratios are the same in both directions (symmetrical stretching). For extremely high draw ratios almost perfect planar orientation is obtained, that is, the chain axis of each individual molecule is parallel to the xy-plane, which is assumed to be the plane of the film (Fig. 25). The chains are distributed at random in this plane and also with respect to a rotation around their own axis c. The transition moment forms an angle θ with c, and the chain coordinate system a, b, c is chosen so that $\mathbf{M}$ falls in the ac-plane as shown in Fig. 25.

The transformation from the (a, b, c) to the (x, y, z) system is given by

$$(M_x, M_y, M_z) = \begin{pmatrix} \cos\varphi\sin\vartheta & \sin\varphi\sin\vartheta & \cos\vartheta \\ -\cos\varphi\cos\vartheta & -\sin\varphi\cos\vartheta & \sin\vartheta \\ \sin\varphi & -\cos\varphi & 0 \end{pmatrix} \begin{pmatrix} M_a \\ M_b \\ M_c \end{pmatrix} \quad (62)$$

while the components M_a, M_b, and M_c are expressed by (27). To obtain the expected absorption intensities A_x, A_y, and A_z we have to integrate the square of the respective components of $\mathbf{M}$ over φ and ϑ. Equations (27) and (62) lead to

$$A_x = A_y = \frac{M^2}{4\pi^2} \int_0^{2\pi} \int_0^{2\pi} (\cos\varphi \sin\vartheta \sin\theta + \cos\vartheta \cos\theta)^2 \, d\varphi \, d\vartheta$$

$$= \frac{M^2}{4} (1 + \cos^2\theta) \tag{63}$$

$$A_z = \frac{M^2}{4\pi^2} \int_0^{2\pi} \int_0^{2\pi} (\sin\varphi \sin\theta)^2 \, d\varphi \, d\vartheta = \frac{M^2}{2} \sin^2\theta$$

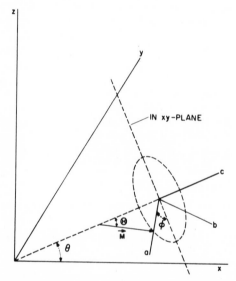

FIG. 25. Perfect symmetrical planar orientation chain axis c in xy-plane.

Since $A_x = A_y$ no dichroism is expected for radiation with a propagation direction perpendicular to the film. Dichroism can only be observed by cutting a cross section and looking at the film edge on or by tilting the sample in the radiation beam as discussed in Section 3D. In this fashion it is possible to determine experimentally a value for $R_{x,z}$ which from (63) is expected to be

$$R_{x,z} = \frac{A_x}{A_z} = \frac{1 + \cos^2\theta}{2\sin^2\theta} \tag{64}$$

This equation is equivalent to (46). It can be used to determine θ.

If tilting and sectioning are not practical there is still a possibility of obtaining information about the planar orientation by using unpolarized radiation normal to the plane of the film. To explain this let us consider a transition moment parallel to the chain axis c ($\theta = 90°$). In a perfect planar oriented sample with all the c-axes in the xy-plane every individual transition moment absorbs very strongly since the electric vector of the radiation is also parallel to this plane. The absorption intensity is expected to be lower for a randomly oriented sample of the same thickness since, for example, a chain with its transition moment parallel to the z-axis does not contribute at all to the absorption.

To describe this behavior quantitatively the intensities in (63) were normalized so that

$$A_x + A_y + A_z = M^2 \tag{65}$$

For a randomly oriented sample the intensities are

$$A_x = A_y = A_z = \tfrac{1}{3}M^2 \tag{66}$$

We now compare two films of the same thickness, one having random the other one perfect symmetric planar orientation. From (63) and (66) we obtain

$$\frac{A_{x,\,\text{planar}}}{A_{x,\,\text{random}}} = \frac{3}{4}(1 + \cos^2\theta) \tag{67}$$

If this intensity ratio can be determined experimentally the angle θ can be computed from (67). Such measurements are possible if the film thickness is accurately known and absolute band intensities can be obtained.

Sometimes it is useful to compare an oriented with an unoriented film of the same polymer and consider the intensity ratio of two absorption bands within the same film. Suppose their respective transition moments form the angles θ_1 and θ_2 with the chain axis. Experimentally we determine the intensity ratios B of the two bands. For an unoriented film B becomes

$$B_{\text{random}} = \frac{A_{x,\,1,\,\text{random}}}{A_{x,\,2,\,\text{random}}} = \frac{M_1^2}{M_2^2} \tag{68}$$

where $A_{x,\,1}$ and $A_{x,\,2}$ are the intensities for the two bands measured with light polarized parallel to the x-axis. Since the intensity ellipsoid is rotational symmetrical B would be the same for unpolarized radiation. From (63) and (68), B for planar orientation becomes

$$B_{\text{planar}} = \frac{A_{x,\,1,\,\text{planar}}}{A_{x,\,2,\,\text{planar}}} = B_{\text{random}} \frac{1 + \cos^2\theta_1}{1 + \cos^2\theta_2} \tag{69}$$

Equation (69) can be used to determine θ_1 if θ_2 is known from some other source of information.

E. PARTIAL SYMMETRICAL PLANAR ORIENTATION (4c in Table V)

The type of orientation discussed in the previous paragraph is a limiting case of partial orientation which now will be considered. The chains are no longer perfectly, but are still preferentially, aligned parallel to the xy-plane and the number of chains pointing in a certain angle interval $\gamma, \gamma + d\gamma$ is determined by a distribution function $f(\gamma)$, where γ is the angle between the chain and the z-axis.

The expressions for the absorption intensities A_x, A_y, and A_z, can be obtained by an integration similar to the one by which Eq. (63) was derived. This specific integration has been carried out already in Section 4C, where a sample with partial axial orientation has been fully analyzed and the dichroic ratio $R_{x,z}$ expressed by Eq. (39). This equation covers the case of symmetrical planar orientation where the parameter S ranges from $-\infty$ for a random to -2 for a perfect symmetrical planar distribution of the molecular chains.

According to (38) and (40) the expected absorption intensities for a rotational symmetrical distribution function are given by

$$A_x = A_y = \frac{M^2}{4}[(2N-3F)\sin^2\theta + 2F]$$

$$A_z = \frac{M^2}{2}[(2N-3F)\cos^2\theta + F]$$

(70)

The distribution function $f(\gamma)$ is usually normalized so that $N = 1$. It is interesting to note that for a perfect planar orientation $F = 1$, and (70) is equivalent to (63).

In analogy to (67) the ratio of the band intensities for a symmetric planar and randomly oriented sample (radiation traveling perpendicular to the plane of the film) is

$$\frac{A_{x,\text{planar}}}{A_{x,\text{random}}} = \frac{3}{4}[(2-3F)\sin^2\theta + 2F]$$

(71)

and in analogy to (69) the ratio B between two bands within the same sample becomes

$$B_{\text{planar}} = B_{\text{random}}\frac{(2-3F)\sin^2\theta_1 + 2F}{(2-3F)\sin^2\theta_2 + 2F}$$

(72)

The discussions of (67) and (69) in the previous subsection apply as well to (71) and (72).

Finally it should be noted that (39), (64), and (67)–(72) also hold for a distribution function which is not rotational symmetrical around the z-axis as long as the intensity ellipsoid has that symmetry, although the orientation parameters S and F are then defined in a slightly different way which will be discussed in the next subsection. Such orientations are shown under 4a II and 4c II of Table V. They can occur in symmetrically two-way stretched films, while in a film drawn homogeneously in all directions (e.g. by a mechanical stretcher with many clamps arranged in a circle and expanding radially) the distribution function would be rotational symmetrical around the z-axis.

F. PREFERRED PLANAR ORIENTATION

General Distribution Function

In this paragraph we will discuss the most general distribution function $f(\gamma, \varphi)$ that still leads to an intensity ellipsoid as discussed in Section 3C. There it was shown that such a function should have a symmetry at least as high as the point group D_{2h}. In the previous subsections the distribution function was either rotational symmetrical or the z-axis was at least a fourfold symmetry axis. For the point group D_{2h} the z-axis is only required to be a twofold symmetry axis.

The expected absorption intensities A_x, A_y, and A_z can again be obtained by equations of the form of (36), where $f(\gamma)$ is replaced by $f(\gamma, \varphi)$:

$$
\begin{aligned}
A_x &= \frac{M^2}{4\pi^2} \int_{\gamma=0}^{\pi/2} \int_0^{2\pi} \int_0^{2\pi} [(\cos\vartheta\cos\gamma\cos\varphi + \sin\vartheta\sin\varphi)\sin\theta \\
&\qquad\qquad\qquad\qquad + \cos\varphi\sin\gamma\cos\theta]^2 f(\gamma,\varphi)\, d\vartheta\, d\varphi\, d\gamma \\
&= \frac{M^2}{4\pi}\left[\sin^2\theta \int_{\gamma=0}^{\pi/2} \int_0^{2\pi} f(\gamma,\varphi)(1 - 3\cos^2\varphi\sin^2\gamma)\, d\varphi\, d\gamma \right. \\
&\qquad\qquad\qquad\left. + \int_{\gamma=0}^{\pi/2} \int_0^{2\pi} f(\gamma,\varphi)\, 2\cos^2\varphi\sin^2\gamma\, d\varphi\, d\gamma \right]
\end{aligned}
$$

$$
\begin{aligned}
A_y &= \frac{M^2}{4\pi^2} \int_{\gamma=0}^{\pi/2} \int_0^{2\pi} \int_0^{2\pi} [(\cos\vartheta\cos\gamma\sin\varphi - \sin\vartheta\cos\varphi)\sin\theta \\
&\qquad\qquad\qquad\qquad - \sin\varphi\sin\gamma\cos\theta]^2 f(\gamma,\varphi)\, d\vartheta\, d\varphi\, d\gamma \quad (73) \\
&= \frac{M^2}{4\pi}\left[\sin^2\theta \int_{\gamma=0}^{\pi/2} \int_0^{2\pi} f(\gamma,\varphi)(1 - 3\sin^2\varphi\sin^2\gamma)\, d\varphi\, d\gamma \right. \\
&\qquad\qquad\qquad\left. + \int_{\gamma=0}^{\pi/2} \int_0^{2\pi} f(\gamma,\varphi)\, 2\sin^2\varphi\sin^2\gamma\, d\varphi\, d\gamma \right]
\end{aligned}
$$

$$
\begin{aligned}
A_z &= \frac{M^2}{4\pi^2} \int_{\gamma=0}^{\pi/2} \int_0^{2\pi} \int_0^{2\pi} [-\sin\gamma\cos\vartheta\sin\theta + \cos\gamma\cos\theta]^2 f(\gamma,\varphi)\, d\vartheta\, d\varphi\, d\gamma \\
&= \frac{M^2}{4\pi}\left[\cos^2\theta \int_{\gamma=0}^{\pi/2} \int_0^{2\pi} f(\gamma,\varphi)(2 - 3\sin^2\gamma)\, d\varphi\, d\gamma \right. \\
&\qquad\qquad\qquad\left. + \int_{\gamma=0}^{\pi/2} \int_0^{2\pi} f(\gamma,\varphi)\sin^2\gamma\, d\varphi\, d\gamma \right]
\end{aligned}
$$

In these integrations we have taken advantage of the fact that we have an intensity ellipsoid. Therefore, all the cross terms in Eq. (17) disappear. Equations (73) determine the intensity ellipsoid for a general distribution function for samples where the chains are oriented at random with respect to a rotation around their own axis. The most general function without this restriction will be considered in subsection F.

In Section 4C, Eq. (39), we have shown that for axial orientation the dichroic ratio can be expressed as a simple function of θ (angle between transition moment and chain axis) and the orientation parameter S. The same is true for a type of orientation with the more general distribution function $f(\gamma, \varphi)$ if the proper dichroic ratio is chosen. Beer (Ref. 7) has shown that the dichroic ratio

$$R_{\frac{1}{2}xy,\,z} = \frac{A_{\frac{1}{2}xy}}{A_z} \tag{74}$$

has the same form as (39), where $A_{\frac{1}{2}xy}$ is defined by

$$A_{\frac{1}{2}xy} = \tfrac{1}{2}(A_x + A_y) \tag{75}$$

provided the distribution function has a symmetry sufficient for an intensity ellipsoid, which means that $A_{\frac{1}{2}xy}$ is the absorption band intensity measured with the electric vector of the radiation at $45°$ to the x- and y-axes.

From (73), (74), and (75) we obtain in analogy to (39)

$$R_{\frac{1}{2}xy,\,z} = \frac{\sin^2\theta + S_1}{2\cos^2\theta + S_1} \tag{76}$$

where

$$S_1 = \frac{F_1}{N_1 - \tfrac{3}{2}F_1}$$

$$F_1 = \int\limits_{\gamma=0}^{\pi/2}\int\limits_0^{2\pi} f(\gamma, \varphi)\sin^2\gamma\,d\varphi\,d\gamma; \qquad N_1 = \int\limits_{\gamma=0}^{\pi/2}\int\limits_0^{2\pi} f(\gamma, \varphi)\,d\varphi\,d\gamma \tag{77}$$

$f(\gamma, \varphi)$ is usually normalized so that $N_1 = 2\pi$.

The discussion in Section 4C of the dichroic ratio formula (39) also applies to (76) especially with respect to its usefulness for the determination of θ, but (76) cannot give any information about the preferred orientation in the xy-plane. This kind of information can only be obtained from the dichroic ratio involving A_x and A_y individually that is from $R_{x,\,y}$.

Perfect Preferred Planar Orientation

All the intensity and dichroic ratio formulas derived in the previous sections are special cases of (73). They could have been obtained just as well by substituting $f(\gamma, \varphi)$ in (73) by the proper distribution function for each specific example.

In this paragraph we will briefly discuss perfect preferred planar orientation, the intensity ellipsoid of which is also a special case of (73) but has not been treated so far. In this type of orientation the chains are all parallel to the xy-plane but alinged preferentially along a given axis, for example, the y-axis (see 4b in Table V). The distribution function $f(\gamma, \varphi)$ reduces to the form

$$f(\gamma, \varphi) = \delta(90° - \gamma) f(\varphi) \tag{78}$$

where δ is defined by (43). For this distribution function equations (73) can be integrated over γ to give

$$A_x = \frac{M^2}{4\pi} \left[\sin^2 \theta \int_0^{2\pi} f(\varphi)(1 - 3\cos^2 \varphi) \, d\varphi + \int_0^{2\pi} f(\varphi) 2\cos^2 \varphi \, d\varphi \right]$$

$$A_y = \frac{M^2}{4\pi} \left[\sin^2 \theta \int_0^{2\pi} f(\varphi)(1 - 3\sin^2 \varphi) \, d\varphi + \int_0^{2\pi} f(\varphi) 2\sin^2 \varphi \, d\varphi \right] \tag{79}$$

$$A_z = \frac{M^2}{4\pi} \sin^2 \theta \int_0^{2\pi} f(\varphi) \, d\varphi$$

The dichroic ratios corresponding to (79) will not be given here, but we will describe how in analogy to Section 4D [Eqs. (67) and (71)] the angle θ can be determined from intensity measurements on films with a known thickness and comparison with measurements on randomly oriented samples of the same thickness. If $f(\varphi)$ is normalized according to (77) A_z from (79) becomes

$$A_{z, \text{planar}} = \frac{M^2}{2} \sin^2 \theta \tag{80}$$

while (66) gives the intensity for a randomly oriented sample. Equations (80) and (66) lead to

$$\frac{A_{z, \text{planar}}}{A_{z, \text{random}}} = \frac{3}{2} \sin^2 \theta \tag{81}$$

If this ratio is determined experimentally θ can be calculated. $A_{z, \text{planar}}$ can only be measured by a rather tedious tilting or a sectioning procedure as described in Section 3D. It is also possible to obtain its value indirectly by measuring $A_{x, \text{planar}}$ and $A_{y, \text{planar}}$ separately and making use of the relation

$$A_{x, \text{planar}} + A_{y, \text{planar}} + A_{z, \text{planar}} = M^2 \tag{82}$$

Partial Preferred Planar Orientation and Two-Way Stretched Film Model (Kratky)

The orientation in axial and symmetrical planar oriented samples can be characterized by one single orientation parameter as far as the infrared spectrum is concerned. The reason for this is that the intensity ellipsoid is rotational symmetrical. In the previous discussion the choice of orientation parameters was quite arbitrary. S, F, r, f, γ_0 [Eqs. (39), (40), (48), (50)] or the draw ratio v [Eq. (52)] can be used to describe the orientation. All these parameters are equivalent in principle; although some of them might be mathematically more convenient, others might be more useful for visualizing the degree of orientation.

For a preferred planar oriented sample two orientation parameters are required for full characterization, since the intensity ellipsoid is no longer rotational symmetrical. In the following discussion we will consider the "two-way stretched film" model, and as orientation parameters we have chosen the draw ratios v_x and v_y in the two stretching directions. The reason for this choice is that preferred planar orientation mainly occurs in two-way stretched films.

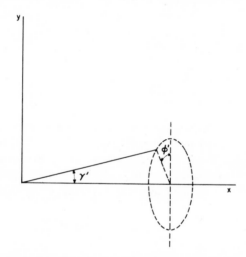

FIG. 26. φ' and γ' are the polar coordinates with respect to the x-axis.

In subsection C, Kratky's model for the axial orientation process has been discussed, which led to a distribution function given by Eq. (57). In the same paper (Ref. 36) using the same model he also derived an equation for the distribution function, $a(\gamma', \varphi')$, obtained by stretching a film in two perpendicular directions. The polar coordinates with respect to the x-axis γ' and φ' are shown in Fig. 26. Kratky's equation is

$$a(\gamma', \varphi') = \frac{v_x^3}{[1 + (v_x^3 - 1)\sin^2\gamma']^{3/2}}$$
$$\times \frac{v_y^3}{\left[1 + (v_y^3 - 1)\left(1 - \cos^2\varphi' \cdot \frac{v_x^3 \sin^2\gamma'}{\cos^2\gamma' + v_x^3 \sin^2\gamma'}\right)\right]^{3/2}} \quad (83)$$

Let us briefly discuss this equation by making the following three remarks:

1. In the case of a one-dimensional stretching process $(v_x = v > 1, v_y = 1)$,

Eq. (83) reduces to (57) if γ' is replaced by γ, since (57) corresponds to an expansion of the sample in the z-direction.

2. The distribution function in the xy-plane is obtained from (83) if $\varphi' = 0$. It can be written in the following form (Ref. 36):

$$a(\gamma', 0) = v_y^{3/2} \frac{(v_x/v_y)^3}{\{1 + [(v_x/v_y)^3 - 1]\sin^2\gamma'\}^{3/2}} \tag{84}$$

If the amount of stretching is the same in both directions ($v_x = v_y > 1$) the distribution function is rotational symmetrical with respect to the z-axis, since $a(\gamma', 0)$ in (84) becomes independent of γ'. This corresponds to a polar diagram shown under 4c, case I in Table V. For actual samples it is possible that the distribution function corresponds to a polar diagram shown in 4c, case III as evidenced by variations in physical properties measured in various directions in the xy-plane or by X-ray or electron diffraction diagrams.

3. Equation (83) is independent of the order in which the stretching operations are carried out but this is not always the case in practice. The orientation behavior might be different for a sample which has been stretched first in the x- and then in the y-direction than for a sample stretched in the reverse order or simultaneously in both directions. Despite these discrepancies the "two-way stretched film" model still seems to be the most realistic one described so far in the literature.

The distribution function $a(\gamma', \varphi')$ is related to $f(\gamma', \varphi')$ by an equation analogous to (34):

$$f(\gamma', \varphi') = a(\gamma', \varphi')\sin\gamma' \tag{85}$$

Since $a(\gamma', \varphi')$ is known explicitly from (83) it is possible to integrate (73) and predict the shape of the intensity ellipsoid for any combination of draw ratios v_x, v_y.

γ and φ are the polar coordinates with respect to the z-axis, while γ' and φ' are the angles with respect to the x-axis. They were chosen for consistency so that the xy-plane coincides with the plane of the film. If we replace γ and φ in (73) by γ' and φ' the absorption intensity components A_x, A_y, and A_z will have to be replaced by A_y, A_z and A_x, respectively. Equation (73) can then be written in the following form:

$$A_y = \frac{M^2}{4\pi}[(N_2 - 3F_2 + 3L_2)\sin^2\theta + 2F_2 - 2L_2]$$

$$A_z = \frac{M^2}{4\pi}[(N_2 - 3L_2)\sin^2\theta + 2L_2] \tag{86}$$

$$A_x = \frac{M^2}{4\pi}[(2N_2 - 3F_2)\cos^2\theta + F_2]$$

where N_2, F_2, and L_2 are defined analogous to (77):

$$N_2 = \int\limits_{\gamma=0}^{\pi/2} \int\limits_{0}^{2\pi} f(\gamma', \varphi')\, d\varphi'\, d\gamma' = 2\pi$$

$$F_2 = \int\limits_{\gamma=0}^{\pi/2} \int\limits_{0}^{2\pi} f(\gamma', \varphi')\sin^2\gamma'\, d\varphi'\, d\gamma' \qquad (87)$$

$$L_2 = \int\limits_{\gamma=0}^{\pi/2} \int\limits_{0}^{2\pi} f(\gamma', \varphi')\sin^2\varphi'\sin^2\gamma'\, d\varphi'\, d\gamma'$$

The expected components of the absorption intensities and, therefore, the dichroic ratios can easily be calculated by (86) if the expressions (87) are known. These integrations are rather tedious. We have, therefore, carried them out numerically on an electronic computer. The values of $F_2/2\pi$ and $L_2/2\pi$ are plotted in Figs. 27a and 27b as a function of the two draw ratios v_x and v_y. These figures can be used directly to obtain the F_2 and L_2 values required in (86).

G. Uniaxial and Uniplanar Orientation

The types of orientation discussed so far could all be fully characterized by a distribution function for the chain segment orientation alone. Two parameters, φ and γ or φ' and γ', were sufficient, since a random distribution of chains with respect to a rotation around their own axis was assumed. In the case of uniaxial and uniplanar orientation this is no longer true, and chains are rotated preferentially in one direction. Mathematically this means that the distribution function now depends on three parameters and has the general form $f(\gamma, \varphi, \vartheta)$. We first will briefly describe in general terms how such orientations might be obtained and then indicate how the expected absorption intensities can be calculated.

Uniplanar and uniaxial orientations can be obtained by various stretching processes, but none of them has yet been analyzed in a quantitative fashion. We will discuss qualitatively how such an orientation process might be visualized.

If the cross sectional shape of an individual chain is rather flat and a film is two way stretched, the chains will not only align parallel to the plane but they will also tend to rotate so that the plane of the chain is parallel to the plane of the film. This effect is even more pronounced if the polymer is partially precrystallized before stretching and if the microcrystallites have a platelet like shape. In such a case the platelets would tend to align parallel to the plane of the film.

In recent work (Refs. 24, 34) it has become apparent that undrawn polymer samples can contain quite large (several microns) single crystals. In some cases they are thin platelets with the chain axes approximately perpendicular to the plane of the platelet. It has been postulated that the chains are folded as shown

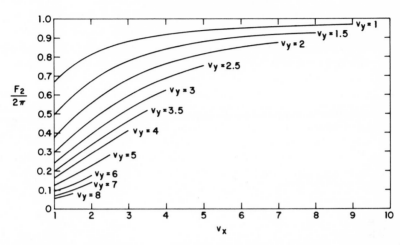

FIG. 27a. Numerical values for the integral F_2 [Eq. (87)] as a function of the draw ratios v_x and v_y.

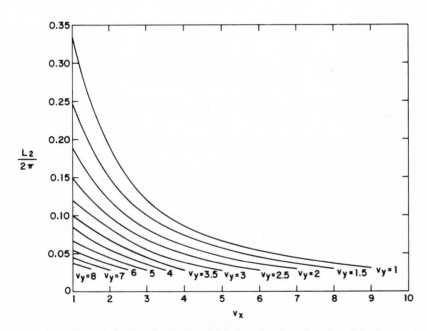

FIG. 27b. Numerical values for the integral L_2 [Eq. (87)] as a function of the draw ratios v_x and v_y.

schematically in Fig. 28. The single crystal platelet usually consists of several layers with an individual thickness h in the range 60–200 Å depending upon the polymer and its physical history. The stretching process of such single crystals has not been studied yet, but it is conceivable that a two-way stretched sample might show some uniplanar orientation if the stretching is visualized as a shearing operation where the single crystals are being deformed.

If ribbon shaped polymer samples are one-way stretched to produce a high axial orientation it can sometimes be noticed that the reduction in width is much less than the reduction in thickness. This is particularly so if the sample is stretched over a heated rod where friction on the rod might restrain the ribbon from reducing appreciably in width. This restriction in motion is to some extent equivalent to a stretching in the plane of the ribbon perpendicular to the main draw direction. It

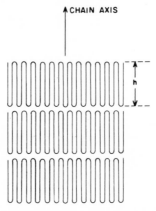

FIG. 28. Schematic diagram of the "folded chain" model for polymer single crystals.

can produce a preferential rotation of the individual chains or crystallites and result in a partial uniaxial orientation.

Perfect uniaxial orientation is the highest degree of orientation possible. All the chains (for an amorphous sample) or all the crystallites (for a microcrystalline polymer) are aligned parallel and show exactly the same orientation with respect to a rotation around the stretching direction. The infrared spectrum is, therefore, equivalent to the one of a single crystal (apart from the edge effects discussed in Chapter IV). This means that all the transition moments of a certain kind are parallel. If we consider a sample coordinate system x, y, z with the polar coordinates γ_0 and φ_0 for the orientation of the transition moments, we obtain for the expected absorption intensities

$$A_x = M^2 \sin^2 \gamma_0 \cos^2 \varphi_0$$
$$A_y = M^2 \sin^2 \gamma_0 \sin^2 \varphi_0 \qquad (88)$$
$$A_z = M^2 \cos^2 \gamma_0$$

In some cases it is possible that a certain functional group, and therefore its transition moment, has several discrete orientations. Under these conditions the absorption intensities (88) will have to be calculated separately for each orientation direction. A simple summation will lead to the total absorption intensity.

Partial uniplanar orientation is the most general type. The expected intensity ellipsoid can be obtained by equation (73) if $f(\gamma, \varphi)$ is replaced by $f(\gamma, \varphi, \vartheta)$ as indicated at the beginning of this subsection. This means that the integration over ϑ in (73) is no longer trivial. It has not yet been carried out for this most general case mainly because there is no model for the orientation process available that would lead to an explicit form for the distribution function $f(\gamma, \varphi, \vartheta)$.

5. Examples

A. LINEAR POLYETHYLENE

Polarization spectra of a thin film of polyethylene, drawn tenfold are shown in Fig. 15b, Chapter II. The strongest absorption bands with their observed dichroism are listed in Table VII. For the following discussion, we will assume that the

TABLE VII

Frequency (cm^{-1})	Type of vibration of the CH$_2$ group	Species (space group)	Observed dichroic ratio $R_{x,z}$ [See Eq. (19c)]
2919	Asym. stretching	$B_{2u} + B_{3u}$	2.54 ± 0.15
2851	Sym. stretching	$B_{2u} + B_{3u}$	2.98 ± 0.15
1473 } 1463 }	Bending	B_{3u} B_{2u}	5.9 ± 0.3
731 } 720 }	Rocking	B_{3u} B_{2u}	8.9 ± 0.9 10.0 ± 0.5

orientation is rotational symmetrical around the z-axis (stretching direction). This would be definitely true for a round fiber, but is probably only an approximation to the actual distribution of crystallites in a film strip.

The transition moments for the vibrations of a CH$_2$ group are quite well defined (see Fig. 8, Chapter IV). They form an angle of exactly 90° with the chain axis for all the vibrations listed in Table VII. We, therefore, would expect to observe dichroic ratios of infinity for a perfectly oriented sample. An X-ray diffraction pattern of the particular film strip showed diffraction spots with a total intensity half-width of about 6° (0° perfect orientation, 180° random distribution of crystallites). This sample is, therefore, highly oriented. If we assume a Kratky type distribution function as shown in Fig. 23, the half-width $\gamma_{1/2}$, defined by (58), is approximately equal to one half the X-ray orientation angle:

$$\gamma_{1/2} = 3°$$

From Fig. 24a we find that this corresponds to a draw ratio of $v = 6$. Since the film was drawn tenfold, we conclude that the orientation efficiency is smaller than expected from Kratky's model. But we nevertheless will assume a Kratky type distribution function because it is still the most realistic one among the ones treated analytically in Section 4C. With $\gamma_{1/2} = 3°$ the orientation parameter S obtained from Fig. 24d is $S = 0.115$. From this parameter and the observed dichroic ratio $R_{x,z}$ we can obtain the angle θ, the transition moment forms with the chain axis by using either Eq. (39) or Fig. 19. The results are given in Table VIII. Only the CH_2-rocking bands behave like perfect perpendicular bands while the transition moment for the bending and stretching vibrations occur at apparent angles θ, substantially smaller than 90°. Bradburry, Elliott, and Frazer (Ref. 15)

TABLE VIII

	$R_{x,z}$	θ (deg)
CH_2-rocking bands	10.0, 8.9	87–90
CH_2-bending bands	5.9	79
CH_2-stretching bands	2.54, 2.98	68–71

and Nielsen and Holland (Refs. 46, 47) have explained this discrepancy by pointing out that amorphous polyethylene has strong absorptions in the 1470 and 2900 cm^{-1} regions while the band in the 720 cm^{-1} region is broad and rather weak (see spectrum of melt, Fig. 15a, Chapter II). The amorphous regions of a sample are usually less oriented than the microcrystallites. One would, therefore, expect a smaller dichroic ratio for the CH_2-bending and stretching bands that are super positions of amorphous and crystalline components. We will now discuss this effect in a quantitative fashion and show that it can only partially account for the rather low dichroism of the CH_2-stretching vibrations. The spectra in Fig. 15b, Chapter II, are those of a highly linear, high density polyethylene sample the crystallinity of which is higher than 80% (see e.g. Refs. 45, 59). If we assume that the amorphous fraction of an 80% crystalline sample is completely unoriented and the crystallites are highly oriented with $\gamma_{1/2} = 3°$ (as determined by X-ray diffraction) the expected dichroic ratio would be 4.8. This is a minimum value since it is very unlikely that in a highly drawn sample the amorphous regions are completely unoriented. From that discussion, it is clear that the dichroic ratio of 5.9 for the 1470 cm^{-1} band can be explained by the presence of a less oriented amorphous fraction, but it can only account partially for the values below three for the CH_2-stretching vibrations. To explain this relatively low dichroism another effect will be discussed in the following paragraph.

Consider the symmetric CH_2-stretching vibration with a transition moment on the bisector of the angle formed by the two C—H bonds. We now note that the CH_2 group is also involved in other vibrations, for example, the CH_2-wagging vibration (Fig. 8, Chapter IV), for which the hydrogen atoms move essentially parallel to the chain axis. The transition moment for the stretching vibration therefore, does not, stay at a fixed angle of $90°$ with respect to the chain, but oscillates within the angle of $90° \pm \delta$. The stretching vibration is fast compared to the wagging vibration. This means that a full cycle of a stretching oscillation may occur while the transition moment forms an angle of essentially $90° - \delta$ with the chain axis, thus reducing the observed dichroic ratio while the dichroic ratio is not reduced by this effect for the low frequency CH_2-rocking vibration since the angle δ is averaged out to zero over one cycle of the rocking vibration.

Let us now consider this effect quantitatively to show that it can indeed account for the observed reduction in dichroism. At room temperature most of the molecules are in the vibrational ground state, the energy of which is

$$E_0 = \tfrac{1}{2}h\nu \tag{89}$$

For an individual CH_2 group, we approximate the wagging vibration by a harmonic oscillator of the same frequency which is about 1300 cm^{-1}. The zero point energy E_0 (oscillator in the ground state) can also be expressed by

$$E_0 = \tfrac{1}{2}k_b l^2 \delta_{\max}^2 \tag{90}$$

Where k_b is the bending force constant, l is the C—H bond length and $\delta_{\max}$ is the maximum angle of the zero point oscillation. It can be calculated by equating the right-hand sides of Eqs. (89) and (90) using the following numerical values:

$$l \sim 10^{-8} \text{ cm}$$
$$k_b \sim 5 \times 10^4 \text{ dynes/cm (Ref. 31)}$$
$$\nu \sim 1300 \text{ cm}^{-1} = 39 \times 10^{12} \text{ sec}^{-1}$$
$$h = 6.626 \times 10^{-27} \text{ erg sec}$$

The result of the computation is

$$\delta_{\max} = 0.227 \text{ or } 13°$$

During the wagging oscillation, the velocity is zero at an angle δ of $+$ and $-13°$. This means that on the average the angle δ is closer to $13°$ than $0°$. For the following discussion, we assume the average angle δ to be $10°$; the angle θ [Eq. (39)] then becomes $80°$. For a value of 0.115 for the orientation parameter S, as determined above, and $\theta = 80°$ the dichroic ratio is 6.1 [Eq. (39) or Fig. 19], assuming no amorphous, unoriented fraction. (The presence of such a fraction would, of course, reduce this value.) The CH_2-wagging vibration is not the only one that contributes to the angle δ. Other parallel vibrations have the same effect. There are also some skeletal bending modes for which the CH_2 group is being bent out of its perpendicular position. All these vibrations combined will tend to increase the average angle δ. If we assume that it really is 20° instead of 10°, as calculated for only one parallel vibration, the dichroic ratio $R_{x,z}$ becomes 2.85 (assuming again a value of 0.115 for S). Such a dichroic ratio is in line with the observed ones for the CH_2-stretching vibrations.

B. Isotactic Polystyrene

Isotactic polystyrene is a stereoregular polymer. In its crystalline form, the carbon backbone chain forms a helix with a crystallographic repeat distance of 6.65 Å. The repeat unit of the helix consists of three monomer units $CH_2CHC_6H_5$.

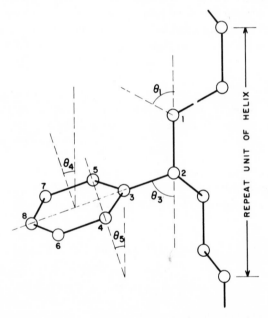

Fig. 29. Helical structure of isotactic, crystalline polystyrene, after Natta, Corradini, and Bassi (Ref. 43). Only one of the three benzene rings per helical repeat unit is shown.

A view of such a helix, perpendicular to its axis, is shown schematically in Fig. 29. A full account of the structure as well as a review of the literature on this subject was given by Natta, Corradini, and Bassi (Ref. 43).

Various authors have published infrared spectra of atactic and isotactic polystyrene (see. e.g., Refs. 39, 42, 55). A summary of band assignments has been given by Krimm (Ref. 37). Most of the observed absorption bands can be assigned to vibrations of the CH_2, CH, or phenyl group. Effects due to intra- or intermolecular interaction are pronounced for a few absorption bands, but in a first approximation the observed spectrum is the one of a single monomer unit.

For the helical model shown in Fig. 29, the relative position of the atoms as well as the angle a certain transition moment forms with the axis of the helix is the same

TABLE IX

DICHROISM OF SOME VIBRATIONS IN THE SPECTRUM OF α-d-POLYSTYRENE

Type of vibration	Assignment (Ref. 37)	Freq. (cm^{-1})	Dichroic ratio $R_{x,z}$ observed	θ for structure given in Ref. 43	θ (calc. from dichroism)
Aromatic CH-stretching vibration (transition moment, perpendicular to axis[a] and in plane of benzene ring)	ν_{20B}	3086	1.76	$\theta_5 = 65°$	68°
Two aromatic CH-stretching vibrations (transition moment parallel to axis[a] of benzene ring)	ν'_{20A}	3065	1.53	$\theta_3 = 71°$	65°
	ν'_2	3026	1.69	$\theta_3 = 71°$	67°
Asym. CH_2-stretching vibration	ν_a	2924	1.39	$\theta_3 = 90°$	63°
Sym. CH_2-stretching vibration	ν_s	2865	0.96	$\theta_1 = 60°$	54°
CD-stretching vibration		2150	~2.1	$\theta_3 = 71°$	72°
Sym. in-plane aromatic CH-bending vibration	ν_{18A}	1027	1.99	$\theta_3 = 71°$	71° assumed as basis for calculations
Aromatic CH out-of-plane bending vibration	ν_{10B}	763	0.373	$\theta_4 = 34°$	30°

[a] Axis of benzene ring is the line connecting carbon atoms 2, 3, and 8 in Fig. 29.

for all monomer units. The direction of a transition moment within an absorbing group is not well known except for some very localized motions such as for CH-stretching vibrations or the CH out-of-plane bending vibrations of the benzene ring. For the latter, the transition moment is perpendicular to the ring. The dichroic behavior of the absorption bands of such vibrations can, therefore, be calculated if the geometrical structure of the chain and the degree of orientation are known. In the following paragraphs, we will compare such calculations with experimental data.

Natta, Corradini, and Bassi (Ref. 43) have given the coordinates of all atoms

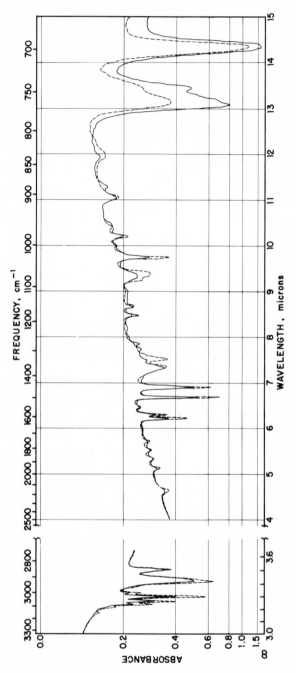

FIG. 30. Infrared spectra of oriented isotactic α-deuterated polystyrene. KEY: ———: electric vector parallel to the direction of stretch; – – –: electric vector perpendicular to the direction of stretch. From 3 to 3.6 μ the spectrum was recorded with a Perkin-Elmer single beam double pass spectrophotometer equipped with a CaF₂ prism. From 4 to 15 μ a Perkin-Elmer model 21 (NaCl prism) was used.

for crystalline polystyrene. From their data, one can calculate the angles θ of certain groups with respect to the axis of the helix. The results of such calculations are given in column 5, Table IX for some characteristic group vibrations with a fairly well defined transition moment within the group. The assignment of the vibrations (column 2) is the same as in Krimm's review article (Ref. 37). Some of the angles θ are also shown in Fig. 29.

Polarized spectra of a stretched film of isotactic α-d-polystyrene are reproduced in Fig. 30. The α-deuterated polymer was used to avoid interference of the CH_2- and CH-stretching absorption bands. Dichroic ratios were measured for eight bands. The observed values are listed in column 4 of Table IX.

The type of orientation was not known for the sample measured in Fig. 30, but since the film was one-way stretched we assume axial orientation with rotational symmetry around the stretching direction, which is the simplest type of orientation consistent with the sample preparation. The expected dichroic ratio for this type of orientation is given by Eq. (39) as a function of the angle θ and a single orientation parameter S. For convenience this function is plotted in Fig. 19. We now choose the 1027 cm^{-1} band (a band with a fairly high dichroic ratio) to determine the value of S. For $R_{x,z} = 1.99$ and $\theta_3 = 71°$ (from model Ref. 43) and the help of Fig. 19 we obtain $S = 0.48$. With this value of S we can calculate the angle θ the transition moment forms with the axis of the helix for the remaining seven absorption bands.† Such calculated values are given in the last column of Table IX. The agreement between the last two columns is quite good except for the antisymmetric CH_2-stretching vibration for which the observed dichroic ratio and, therefore, the calculated θ are much too small. There are two reasons that might be responsible for this discrepancy. First, the real structure of the molecule may deviate somewhat from the one given in Ref. 43, that is, the CH_2 group may be somewhat twisted so that the transition moment of the antisymmetric stretching vibration would not be exactly perpendicular to the axis of the helix. It is not likely that such a deviation is large enough to be the sole cause of the large discrepancy in dichroism. The second effect, discussed in detail in the previous subsection in connection with polyethylene, is probably also contributing to the lowering of the dichroic ratio. In this effect other vibrations such as the CH_2-twisting vibration cause the average angle θ of the antisymmetric CH_2-stretching vibration to be smaller than 90°. The symmetric CH_2-stretching vibration, on the other hand, is not affected very much since for its transition moment θ is about 54 to 60° and the dichroic ratio about one. A deviation from this angle by $\Delta\theta$ due to a superimposed lower frequency vibration would during one phase of the vibration $(\theta - \Delta\theta)$ lead to a dichroic ratio smaller than one and during the other phase $(\theta + \Delta\theta)$ to a ratio larger than one and on the average a ratio of about one would still be observed.

† We could also have chosen another band, for example, the 763 cm^{-1} band and assume the angle $\theta_4 = 34°$ as a basis for the determination of S and the other angles θ. The numerical values for S and θ would have been slightly but not significantly different.

REFERENCES

1. Abbott, N. B., and Elliott, A., *Proc. Roy. Soc.* **A234**, 247 (1956).
2. Anderson, D. H., and Miller, O. E., *J. Opt. Soc. Am.* **43**, 777 (1953).
3. Badger, R. M., and Newman, R., *Rev. Sci. Instr.* **22**, 935 (1951).
4. Barer, R., Cole, A. R. H., and Thompson, H. W., *Nature* **163**, 198 (1949).
5. Bauman, R. P., "Advanced Analytical Chemistry." McGraw-Hill, New York, 1958.
6. Beer, M., private communication, 1955.
7. Beer, M., *Proc. Roy. Soc.* **A236**, 136 (1956).
8. Beste, L. F., private communication, 1959.
9. Bird, G. R., and Shurcliff, W. A., *J. Opt. Soc. Am.* **49**, 235 (1959).
10. Blout, E. R., *in* "Techniques of Organic Chemistry," (A. Weissberger, ed.), Vol. 1, Part 2 (3rd ed.), p. 1475. Wiley (Interscience), New York, 1960.
11. Blout, E. R., and Bird, G. R., *J. Opt. Soc. Am.* **41**, 547 (1951).
12. Blout, E. R., Bird, G. R., and Grey, D. S., *J. Opt. Soc. Am.* **40**, 304 (1950).
13. Bohn, C. R., private communication, 1956.
14. Born, M., "Optik." Springer, Berlin, 1933.
15. Bradbury, E. M., Elliott, A., and Fraser, R. D. B., *Trans. Faraday Soc.* **56**, 1117 (1960).
16. Cannon, C. G., *Chem. & Ind.* (*London*) p. 29 (1957).
17. Cannon, C. G., *in* "Physical Methods of Investigating Textiles" (R. Meredith and J. W. S. Hearle, eds.). Wiley (Interscience), New York, 1959.
18. Cloud, W. H., *J. Opt. Soc. Am.* **46**, 899 (1956).
19. Coates, V. J., Offner, A., and Siegler, E. H., *J. Opt. Soc. Am.* **43**, 984 (1953).
20. Cole, A. R. H., and Jones, R. N., *J. Opt. Soc. Am.* **42**, 348 (1952).
21. Edwards, D. F., and Bruemmer, M. J., *J. Opt. Soc. Am.* **49**, 860 (1959).
22. Elliott, A., Ambrose, E. J., and Temple, R. B., *J. Opt. Soc. Am.* **38**, 212 (1948).
23. Elliott, A., Ambrose, E. J., and Temple, R. B., *J. Sci. Instr.* **27**, 21 (1950).
24. Fischer, E. W., *Z. Naturforsch.* **12a**, 753 (1957).
25. Fraser, R. D. B., *Discussions Faraday Soc.* **9**, 378 (1950).
26. Fraser, R. D. B., *J. Chem. Phys.* **21**, 1511 (1953).
27. Fraser, R. D. B., *J. Opt. Soc. Am.* **43**, 929 (1953).
28. Fraser, R. D. B., *J. Chem. Phys.* **28**, 1113 (1958).
29. Fraser, R. D. B., *J. Chem. Phys.* **29**, 1428 (1958).
30. Gore, R., *Science*, **110**, 710 (1949).
31. Herzberg, G., "Infrared and Raman Spectra of Polyatomic Molecules." Van Nostrand, Princeton, New Jersey, 1945.
32. Jones, A. V., Ph.D. Thesis, Cambridge University, 1949.
33. Jones, R. C., *J. Opt. Soc. Am.* **46**, 528 (1956).
34. Keller, A., *Phil. Mag.* [8] **2**, 1171 (1957).
35. Keller, A., and Sandeman, I., *J. Polymer Sci.* **15**, 133 (1955).
36. Kratky, O., *Kolloid-Z.* **64**, 213 (1933).
37. Krimm, S., *Fortschr. Hochpolymer. Forsch.* **2**, 51 (1960).
38. Kuhn, W., and Grün, F., *Kolloid-Z.* **101**, 248 (1942).
39. Liang, C. Y., and Krimm, S., *J. Polymer Sci.* **27**, 241 (1958).
40. Mann, J., and Phil, D., *Skinner's Silk & Rayon Record* **30**, 1262 (1956).
41. Meier, R., and Günthard, H. H., *J. Opt. Soc. Am.* **49**, 1122 (1959).
42. Morero, D., Mantica, E., Ciampelli, F., and Sianesi, D., *Nuovo Cimento* **15**, Suppl., 122 (1960).
43. Natta, G., Corradini, P., and Bassi, I. W., *Nuovo Cimento* **15**, Suppl., 68 (1960).
44. Newman, R., and Halford, R. S., *Rev. Sci. Instr.* **19**, 270 (1948).

45. Nichols, J. B., *J. Appl. Phys.* **25**, 840 (1954).
46. Nielsen, J. R., and Holland, R. F., *J. Mol. Spectry.* **4**, 488 (1960).
47. Nielsen, J. R., and Holland, R. F., *J. Mol. Spectry.* **6**, 394 (1961).
48. Pfund, A. H., *Astrophys. J.* **24**, 19 (1906).
49. Quynn, R. G., Ph.D. Thesis, Princeton University, 1957.
50. Ramsay, D. A., *J. Am. Chem. Soc.* **74**, 72 (1952).
51. Rupert, C. S., *J. Opt. Soc. Am.* **42**, 684 (1952).
52. Sandeman, I., *Proc. Roy. Soc.* **A232**, 105 (1955).
53. Stein, R. S., *J. Polymer. Sci.* **31**, 327 (1958).
54. Stein, R. S., *J. Appl. Polymer Sci.* **5**, 96 (1961).
55. Tadokoro, H., Nishiyama, N., Nozakura, S., and Murahashi, S., *J. Polymer Sci.* **36**, 553 (1959).
56. Tink, R. R., private communication, 1958.
57. Tobin, M. C., and Carrano, M. J., *J. Polymer Sci.* **24**, 93 (1957).
58. Wood, D. L., *Rev. Sci. Instr.* **21**, 764 (1950).
59. Wood, D. L., and Luongo, J. P., *Mod. Plastics* **38**, 132 (1961).
60. Wright, N., *J. Opt. Soc. Am.* **38**, 69 (1948).

—Appendix—

Guide to the Literature of Individual Polymers

Following is a survey of literature references to infrared spectra of individual polymers. The numbers refer to the alphabetical list of references at the end of the appendix. Polymer classes and individual polymers are arranged in an arbitrary way. For most polymers the references are subdivided in the following fashion:

A. General discussion of spectrum, or extensive band assignments, or vibrational analysis

B. Discussion of specific absorption bands or vibrations

C. Orientation effects, polarization measurements

D. Crystallinity effects and measurements

E. Analytical measurements

A reference with a general discussion of the spectrum listed under A may also contain more specific information without being listed under B–E.

Recent review articles on IR spectra of high polymers were written by Schnell (Ref. 247), Elliott (Ref. 55), and Krimm (Ref. 127). Hummel's book (Ref. 44, Chapter I) is very useful for analytical work.

$-(-CH_2-)-_x$ Chains

---

1. Polyethylene $-(-CH_2-CH_2-)-_x$

A   31, 77, 91, 118, 127, 133, 156, 188, 210, 211, 247, 252, 253, 306, 308, 311, 327

B   71, 116, 120, 125, 186, 209, 218, 275, 284, 332

C   2, 5, 23, 28, 56, 70, 93, 109, 115, 117, 243, 270, 271, 272, 273, 275, 309, 312

D   60, 93, 108, 208, 239, 240, 273, 275, 279, 312, 326, 330

E   21, 32, 87, 232, 240, 326

2. Paraffines, fatty acids, and related compounds

A   7, 23, 27, 29, 30, 40, 45, 107, 112, 170, 230, 235, 236, 252, 259, 264, 265, 282, 311, 313, 327

B   39, 111, 125, 164, 207, 229, 263, 269, 274, 275, 284, 286, 287, 332

C   43, 60, 64, 103, 111

D   60, 230, 281, 285

E   169, 232, 280, 324

Vinyl Polymers $-(-CH_2-CHR-)-_x$

---

3. Polypropylene $-(-CH_2-CH-)-_x$
$\qquad\qquad\qquad\qquad\qquad\quad |$
$\qquad\qquad\qquad\qquad\qquad\; CH_3$

A   65, 78, 79, 98, 127, 145, 146, 147, 153, 155, 168, 227, 228, 247, 291, 310

D   1, 88, 157, 200, 222, 238

E, isotacticity   25, 88, 157, 200, 203, 222, 234, 238, 262

4. Poly(1-butene) $-(-CH_2-CH-)-_x$
$\qquad\qquad\qquad\qquad\qquad\qquad |$
$\qquad\qquad\qquad\qquad\qquad\;\; CH_2-CH_3$

A   291

5. Polystyrene

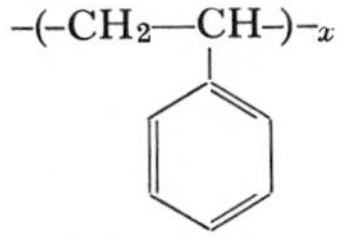

A, atactic   120, 127, 142, 224, 260, 306, 307

isotactic   127, 190, 223, 247, 290, 296, 297

B   51, 128, 298

C   75

D   26

E, copolymers   86, 119, 201, 303, 304, 316

6. Poly(*p*-methyl styrene)

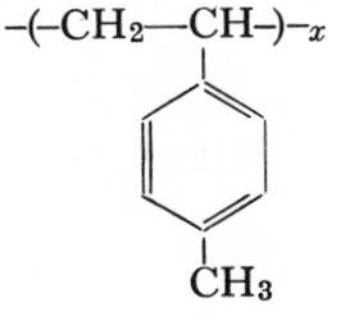

B   298

7. Poly(*m*-methyl styrene)

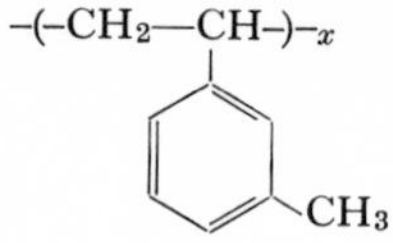

$$-(-CH_2-CH-)-_x$$

   B   298

8. Poly(*p*-fluorostyrene)

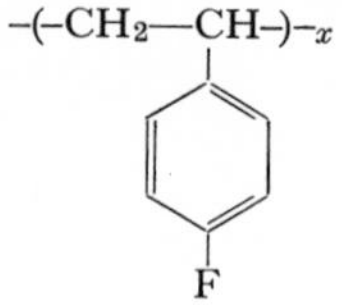

$$-(-CH_2-CH-)-_x$$

   C   123

9. Poly(vinyl cyclopropane)

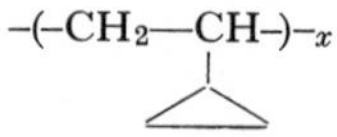

$$-(-CH_2-CH-)-_x$$

   A   206

10. Poly(vinyl cyclohexane)

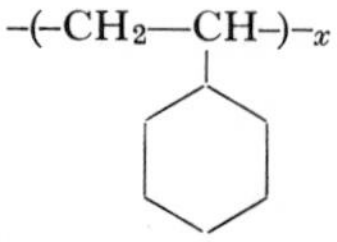

$$-(-CH_2-CH-)-_x$$

   A   291

11. Polyacrolein

$$-(-CH_2-CH-)-_x$$
$$C=O$$
$$H$$

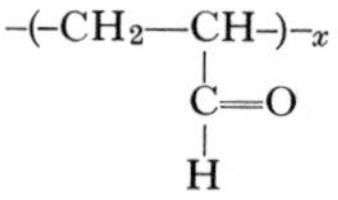

   E, copolymers   249

12. Poly(vinyl alcohol)

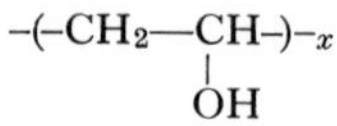

$$-(-CH_2-CH-)-_x$$
$$OH$$

   A   5, 18, 126, 127, 134, 152, 194, 195, 247, 288, 289, 295, 299, 300, 301, 307
   B   72, 73, 83, 165, 294
   C   56, 85, 141, 212, 237
   D   34
   E, branching   99

13. Poly(vinyl acetate)

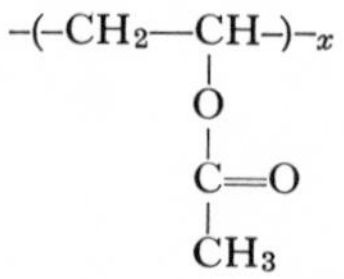

$$-(-CH_2-CH-)-_x$$
$$O$$
$$C=O$$
$$CH_3$$

   A   259, 307
   C   56, 212
   E, branching   99

14. Poly(vinyl trifluoroacetate)

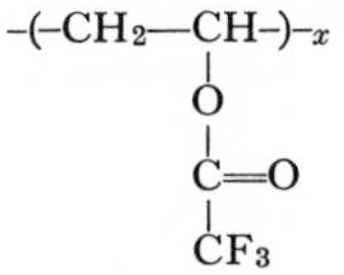

$$-(CH_2-CH-)-_x$$
$$| $$
$$O$$
$$| $$
$$C=O$$
$$| $$
$$CF_3$$

A   84
B   20

15. Poly(vinyl isobutylether)

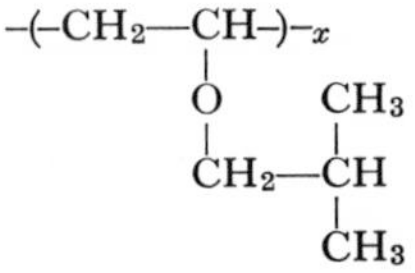

$$-(CH_2-CH-)-_x$$
$$| \quad\quad CH_3$$
$$O \quad\quad |$$
$$CH_2-CH$$
$$| $$
$$CH_3$$

A   202

16. Poly(acrylic acid)

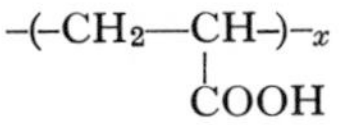

$$-(CH_2-CH-)-_x$$
$$| $$
$$COOH$$

A   250, 259

17. Polyacrylamide

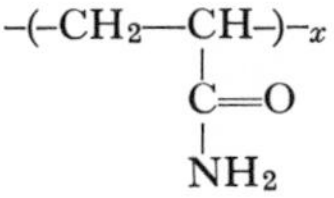

$$-(CH_2-CH-)-_x$$
$$| $$
$$C=O$$
$$| $$
$$NH_2$$

A   250

18. Polyacrylonitrile

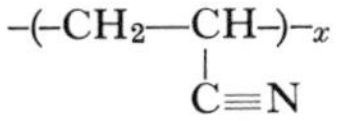

$$-(CH_2-CH-)-_x$$
$$| $$
$$C\equiv N$$

A   8, 143, 154, 250
B   76
C   19, 20
E, copolymers, endgroups   13, 69, 172, 178, 328

19. Poly(vinyl chloride)

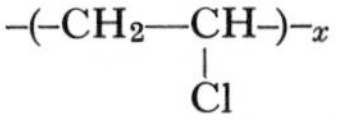

$$-(CH_2-CH-)-_x$$
$$| $$
$$Cl$$

A   105, 120, 126, 127, 199, 250, 254, 305, 306
B   76, 81, 82, 92, 97, 128, 129, 130, 131, 196, 255, 256, 257
C   5, 56
D   114

20. Poly(vinyl nitrate)

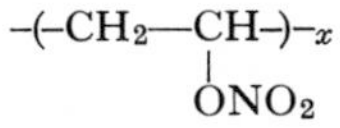

$$-(CH_2-CH-)-_x$$
$$| $$
$$ONO_2$$

A   132

21. Vinyl polymers containing silicon

A   11, 167

Vinylidene Polymers $-(-CH_2-CR_1R_2-)-_x$

22. Polyisobutylene

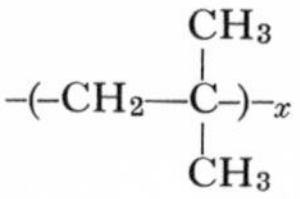

$$-(-CH_2-\overset{\displaystyle CH_3}{\underset{\displaystyle CH_3}{C}}-)-_x$$

   A   95, 100, 247

23. Poly(methyl methacrylate)

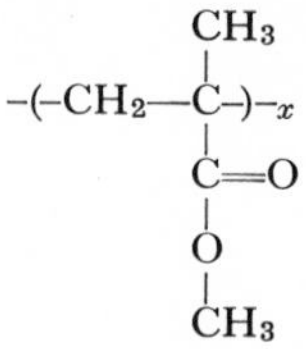

$$-(-CH_2-\overset{\displaystyle CH_3}{\underset{\displaystyle \underset{\displaystyle \underset{\displaystyle CH_3}{O}}{C=O}}{C}}-)-_x$$

   A, atactic   12, 110, 307
       stereoregular   12, 110, 231

24. Various polymethacrylates and polychloromethacrylates

   A   244

25. Polymethacrylonitrile

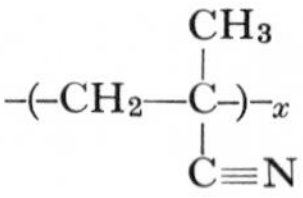

$$-(-CH_2-\overset{\displaystyle CH_3}{\underset{\displaystyle C\equiv N}{C}}-)-_x$$

   A   250, 261

26. Poly(vinylidene fluoride)

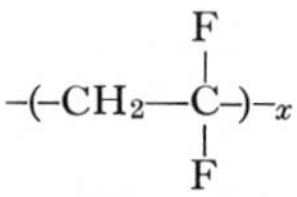

$$-(-CH_2-\overset{\displaystyle F}{\underset{\displaystyle F}{C}}-)-_x$$

   E, decomposition   323

27. Poly(vinylidene chloride)

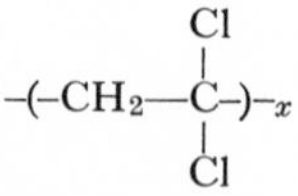

$$-(-CH_2-\overset{\displaystyle Cl}{\underset{\displaystyle Cl}{C}}-)-_x$$

   A   105, 131, 196, 197, 198, 247
   B   76

28. Poly(vinylidene bromide)

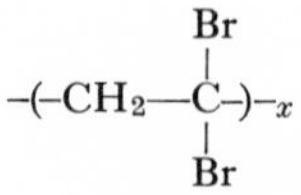

$$-(-CH_2-\overset{\displaystyle Br}{\underset{\displaystyle Br}{C}}-)-_x$$

   A   198

Polymers of the Type $-(-\underset{R_2}{\overset{R_1}{C}}-\underset{R_4}{\overset{R_3}{C}}-)_x$

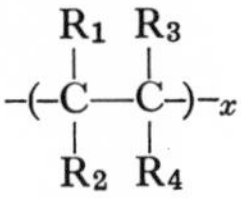

---

29. Polytetrafluoroethylene $-(-\underset{F}{\overset{F}{C}}-\underset{F}{\overset{F}{C}}-)_x$

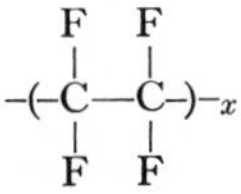

A   127, 139, 193, 247
B   188, 242
D   171, 233
E, copolymers   102

30. Polychlorotrifluoroethylene $-(-\underset{Cl}{\overset{F}{C}}-\underset{F}{\overset{F}{C}}-)_x$

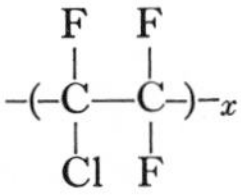

A   34, 139, 247, 322
D   166, 233

infrared optical properties   104

Other Polymers of Unsaturated Monomers

---

31. Polyhydroxymethylene $-(-\underset{OH}{\overset{}{CH}}-\underset{OH}{\overset{}{CH}}-)_x$

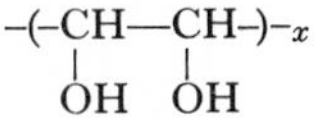

A   63

32. Polyacetoxymethylene $-(-\underset{\underset{\underset{CH_3}{|}}{\overset{|}{C=O}}}{\overset{|}{O}}CH-\underset{\underset{\underset{CH_3}{|}}{\overset{|}{C=O}}}{\overset{|}{O}}CH-)_x$

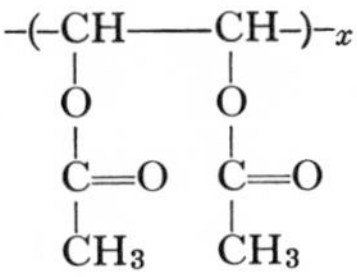

A   63

33. Poly(vinylene carbonate) $-(-CH-CH-)_x$ with $O-\underset{\overset{\|}{O}}{C}-O$ bridge

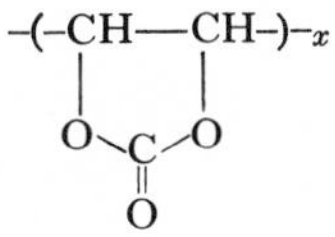

A   63

34. Polybutadiene $-(-CH_2-CH=CH-CH_2-)-_x$ $-(-CH_2-CH-)-_x$
                                                           $CH=CH_2$

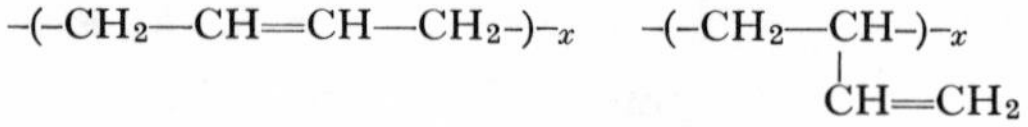

     A   74, 191, 247
     B   62, 192
     E   15, 61, 86

35. Poly(2,3-dimethylbutadiene) $-(-CH_2-C=C-CH_2-)-_x$
                                               CH_3 CH_3

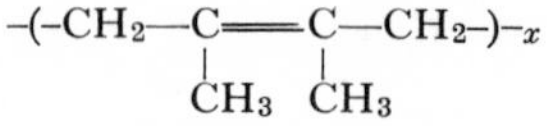

     A   329

36. Polypentadiene $-(-CH_2-CH=CH-CH-)-_x$
                                               CH_3

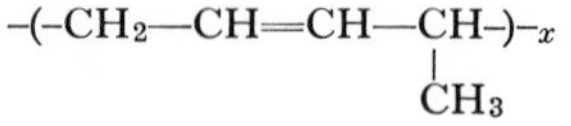

     A   204, 205

37. Polyisoprene

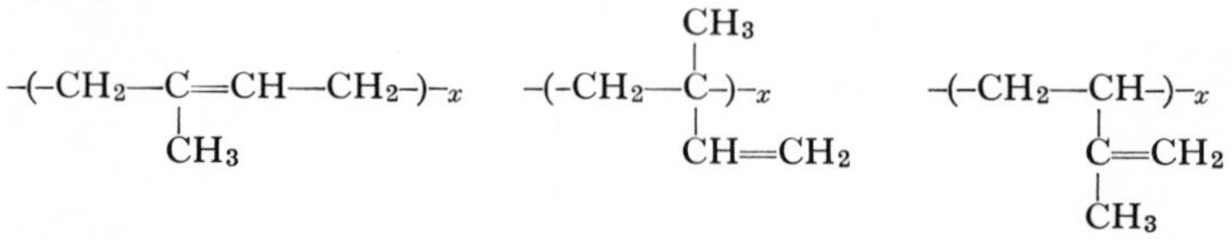

                                      CH_3
$-(-CH_2-C=CH-CH_2-)-_x$ $-(-CH_2-C-)-_x$ $-(-CH_2-CH-)-_x$
       CH_3 $CH=CH_2$ $C=CH_2$
                                                                         CH_3

     B, characterization of various isomers   16, 44, 47, 50, 62, 94, 241, 247, 268
     C   283
     E   258

38. Natural rubber $-(-CH_2 \diagdown C=C \diagup CH_2-)-_x$
                                  $H_3C \diagup$ $\diagdown H$

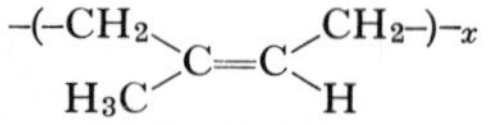

     A   17, 31, 33, 96, 105, 247, 306
     B   50, 120, 318
     C   162
     E   245

39. Polychloroprene $-(-CH_2-CH=C-CH_2-)-_x$
                                           Cl

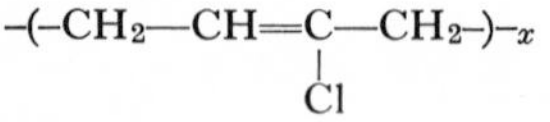

     A   247
     D   208

Polymers of the Form $-(-A-O-)-_x$

40. Polyoxymethylene $-(-CH_2-O-)-_x$
     A   216, 247, 292, 293
     B   186, 187, 188

41. Polyaldehydes $-(-\underset{\underset{\textstyle R}{|}}{CH}-O-)-_x$

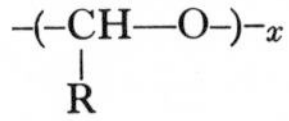

   A   213, 214, 215, 217

42. Poly(ethylene oxide) $-(-CH_2-CH_2-O-)-_x$

   A   49

   B   135, 136, 184, 187, 325

43. Poly(propylene oxide) $-(-CH_2-\underset{\underset{\textstyle CH_3}{|}}{CH}-O-)-_x$

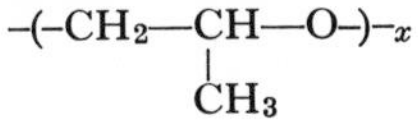

   A   113

44. Substituted poly(ethylene oxides)

   A   101

45. Poly(butadiene monoxide) $-(-CH_2-\underset{\underset{\textstyle CH=CH_2}{|}}{CH}-O-)-_x$

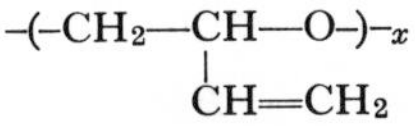

   A   113

46. Poly(styrene oxide) $-(-CH_2-CH-O-)-_x$ (with phenyl substituent)

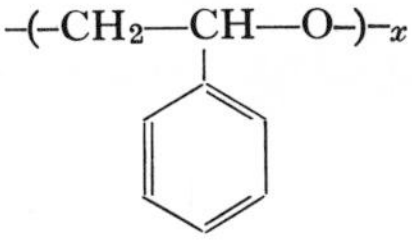

   A   113

47. Polysiloxanes $-(-\underset{\underset{\textstyle R_2}{|}}{\overset{\overset{\textstyle R_1}{|}}{Si}}-O-)-_x$

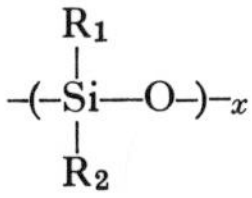

   A   6, 124

Polyesters

---

48. Poly(ethylene terephthalate) $-(-CH_2-CH_2-O-\underset{\underset{\textstyle O}{\|}}{C}-\langle\text{aryl}\rangle-\underset{\underset{\textstyle O}{\|}}{C}-O-)-_x$

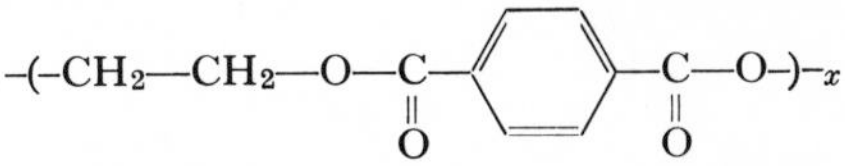

   A   48, 105, 122, 126, 127, 144, 247, 251, 302, 309, 331

   B   128, 174, 175, 176, 182, 183, 319, 320

   C   52, 140, 141, 237

   D   34, 42, 59, 80, 171, 173, 179, 233, 330

   E   225, 321

49. Poly(methylene terephthalate) $-(-CH_2-O-C-\langle\!\!\langle\ \rangle\!\!\rangle-C-O-)-_x$
                                            $\quad\quad\quad\quad\quad\;\; O \quad\quad\quad\quad O$

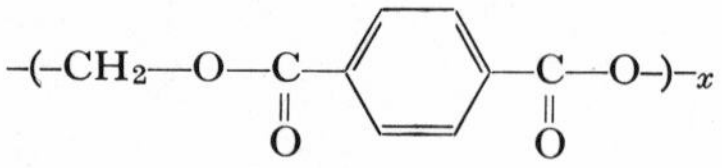

A   58

50. Poly(1,4-cyclohexylene dimethylene terephthalate)

$-(-CH_2-\langle\ \rangle-CH_2-O-C-\langle\!\!\langle\ \rangle\!\!\rangle-C-O-)-_x$
$\quad\quad\quad\quad\quad\quad\quad\quad\quad\quad O \quad\quad\quad\quad O$

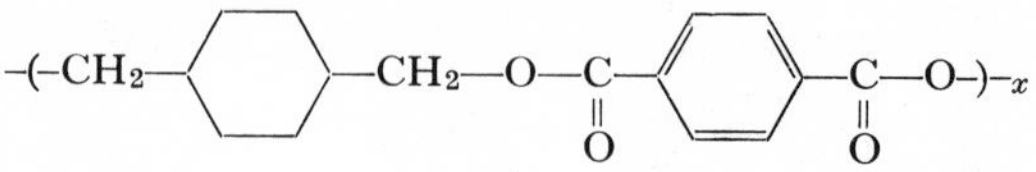

A   22

51. Other polyesters

A   41

Polyamides

52. 6-6 Nylon
$\quad\quad\quad\quad\quad\quad\quad\quad\quad\quad O \quad\quad\quad O$
$-(-NH-[CH_2]_6-NH-C-[CH_2]_4-C-)-_x$

A   35, 177, 247
B   68, 71, 137
C   5, 36, 37, 38, 56, 237
D   208, 267

53. 6-10 Nylon
$\quad\quad\quad\quad\quad\quad\quad\quad\quad\quad O \quad\quad\quad O$
$-(-NH-[CH_2]_6-NH-C-[CH_2]_8-C-)-_x$

A   35
D   504

54. 6 Nylon
$\quad\quad\quad\quad\quad\quad\quad O$
$-(-[CH_2]_5-NH-C-)-_x$

A   35, 247, 311
D   317

55. 11 Nylon
$\quad\quad\quad\quad\quad\quad\quad O$
$-(-[CH_2]_{10}-NH-C-)-_x$

A   311

56. Other nylons

A   31, 35, 121, 177, 180

57. Aromatic polyamides

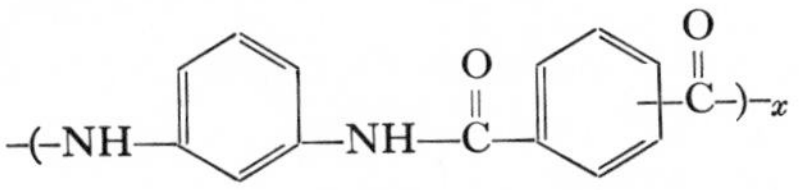

$$-\!(-NH\!-\!\!\bigcirc\!\!-\!NH\!-\!\overset{\overset{\textstyle O}{\|}}{C}\!-\!\!\bigcirc\!\!-\!\overset{\overset{\textstyle O}{\|}}{C}\!-\!)_{-x}$$

  A   161

58. Polypeptides $-\!(-\!\underset{\underset{\textstyle R}{|}}{CH}\!-\!NH\!-\!\underset{\underset{\textstyle O}{\|}}{C}\!-\!)_{-x}$

  A   9, 10, 54, 55 (literature survey up to 1959), 67, 185, 189

  B, α and β helix   53, 57, 138, 266

  C   315

59. Proteins (silk, wool, etc.)

  A   9, 14, 24, 54, 55, 67, 278

  B   3, 4, 68

  C   66, 105

Miscellaneous Polymers

---

60. Poly(p-xylene) $-\!(-CH_2\!-\!\!\bigcirc\!\!-\!CH_2\!-\!)_{-x}$

  A   46, 309

61. Polyphenyls $-\!(-\!\bigcirc\!-\!)_{-x}$

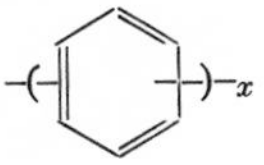

  A   277

62. Polyanhydrides $-\!(-\overset{\overset{\textstyle O}{\|}}{C}\!-\!O\!-\!\overset{\overset{\textstyle O}{\|}}{C}\!-\!A\!-\!)_{-x}$

  A   181

63. Polycarbonates $-\!(-O\!-\!\overset{\overset{\textstyle O}{\|}}{C}\!-\!O\!-\!A\!-\!)_{-x}$

  A   248

64. Cellulose and other polysaccharides

  A   89, 90, 150, 159, 160, 220, 226

  B   55, 148, 149, 158, 163

  C   151, 314

  D   106

  E   219, 221, 276

REFERENCES

1. Abe, K., and Yanagisawa, K., *J. Polymer Sci.* **36**, 536 (1959).
2. Aggarwal, S. L., Tilley, G. P., and Sweeting, O. J., *J. Appl. Polymer Sci.* **1**, 91 (1959).
3. Ambrose, E. J., and Elliott, A., *Textile Res. J.* **22**, 783 (1952).
4. Ambrose, E. J., Elliott, A., and Temple, R. B., *Nature* **163**, 859 (1949).
5. Ambrose, E. J., Elliott, A., and Temple, R. B., *Proc. Roy. Soc.* **A199**, 183 (1949).
6. Andrianov, K. A., Gashnikova, N. P., and Asnovich, E. Z., *Bull. Acad. Sci. U.S.S.R., Div. Chem. Sci. (English Transl.)* p. 800 (1960).
7. Aronovic, S. M., Ph.D. Thesis, University of Wisconsin, 1957.
8. Arthur, J. C., and Demint, R. J., *J. Phys. Chem.* **64**, 1332 (1960).
9. Asai, M., Tsuboi, M., Shimanouchi, T., and Mizushima, S., *J. Phys. Chem.* **59**, 322 (1955).
10. Bamford, C. H., Hanby, W. E., and Elliott, A. E., "Synthetic Polypeptides." Academic Press, New York, 1956.
11. Bassi, I. W., Natta, G., and Corradini, P., *Angew. Chem.* **70**, 597 (1958).
12. Baumann, U., Schreiber, H., and Tessmar, K., *Makromol. Chem.* **36**, 81 (1959).
13. Bayzer, H., and Schurz, J., *Z. Physik. Chem. (Frankfurt)* [N.S.] **13**, 30 (1957).
14. Beer, M., Sutherland, G. B. B. M., Tanner, K. N., and Wood, D. L., *Proc. Roy. Soc.* **A249**, 147 (1959).
15. Binder, J. L., *Anal. Chem.* **26**, 1877 (1954).
16. Binder, J. L., and Ransaw, H. C., *Anal. Chem.* **29**, 503 (1957).
17. Blokh, G. A., and Mal'nev, A. F., *Rubber Chem. Technol.* **32**, 628 (1959).
18. Blout, E. R., and Karplus, R., *J. Am. Chem. Soc.* **70**, 862 (1948).
19. Bohn, C. R., private communication, 1956 (see Chapter V).
20. Bohn, C. R., Schaefgen, J. R., and Statton, W. O., *J. Polymer Sci.* **55**, 531 (1961).
21. Boyd, D. R. J., Voter, R. C., and Bryant, W. M. D., Paper presented at 132nd meeting of the American Chemical Society, New York, 1957. Abstracts p. 8T
22. Boye, C. A., *J. Polymer Sci.* **55**, 263 (1961).
23. Bradbury, E. M., Elliott, A., and Frazer, R. D. B., *Trans. Faraday Soc.* **56**, 1117 (1960).
24. Bradbury, E. M., Price, W. C., and Wilkinson, G. R., *J. Mol. Biol.* **3**, 301 (1961).
25. Brader, J. J., *J. Appl. Polymer Sci.* **3**, 370 (1960).
26. Braun, D., Betz, W., and Kern, W., *Naturwissenschaften* **46**, 444 (1959).
27. Brini, M., *Bull. Soc. Chim. France* p. 996 (1955).
28. Brown, A., *J. Appl. Phys.* **20**, 552 (1949).
29. Brown, J. K., Sheppard, N., and Simpson, D. M., *Discussions Faraday Soc.* **9**, 261 (1950).
30. Brown, J. K., Sheppard, N., and Simpson, D. M., *Phil. Trans. Roy. Soc.* **A247**, 35 (1954).
31. Brügel, W. *Kunststoffe* **46**, 47 (1956).
32. Bryant, W. M. D., and Voter, R. C., *J. Am. Chem. Soc.* **75**, 6113 (1953).
33. Bunn, C. W., *Proc. Roy. Soc.* **A180**, 40 (1942).
34. Bunn, C. W., *J. Appl. Phys.* **25**, 820 (1954).
35. Cannon, C. G., *Spectrochim. Acta* **16**, 302 (1960).
36. Caroti, G., and Dusenbury, J. H., "Effect of Drawing on IR-Dichroism of 6-6 Nylon Filaments." Tech. Rept. No. 16. Textile Res. Inst., Princeton, New Jersey, April 30, 1956.
37. Caroti, G., and Dusenbury, J. H., *Nature* **178**, 162 (1956).
38. Caroti, G., and Dusenbury, J. H., *J. Polymer Sci.* **22**, 399 (1956).
39. Chapman, D., *J. Chem. Soc.* p. 4489 (1957).
40. Chapman, D., *J. Chem. Soc.* p. 784 (1958).

41. Chiang, M., and Bobalek, E. G., *Offic. Dig., Federation Paint Varnish Prod. Clubs* **31**, 1287 (1959).
42. Cobbs, W. H., and Burton, R. L., *J. Polymer Sci.* **10**, 275 (1953).
43. Cole, A. R. H., and Jones, R. N., *J. Opt. Soc. Am.* **42**, 348 (1952).
44. Corish, P. J., *Spectrochim. Acta* **15**, 598 (1959).
45. Corish, P. J., and Chapman, D., *J. Chem. Soc.* p. 1746 (1957).
46. Corley, R. S., Haas, H. C., Kane, M. W., and Livingstone, D. I., *J. Polymer Sci.* **13**, 137 (1954).
47. Cunneen, J. I., Higgins, G. M., and Watson, W. F., *J. Polymer Sci.* **40**, 1 (1959).
48. Daniels, W. W., and Kitson, R. E., *J. Polymer Sci.* **33**, 161 (1958).
49. Davison, W. H. T., *J. Chem. Soc.* p. 3270 (1955).
50. Davison, W. H. T., *Chem. & Ind. (London)* p. 131 (1957).
51. Decamps, E., and Hadni, A., *Compt. Rend.* **250**, 1827 (1960).
52. Dulmage, W. J., and Geddes, A. L., *J. Polymer Sci.* **31**, 499 (1958).
53. Elliott, A., *Proc. Roy. Soc.* **A226**, 408 (1954).
54. Elliott, A., *J. Appl. Chem. (London)* **6**, 341 (1956).
55. Elliott, A., *Advan. Spectry.* **1**, 213 (1959).
56. Elliott, A., Ambrose, E. J., and Temple, R. B., *J. Chem. Phys.* **16**, 877 (1948).
57. Elliott, A., Hanby, W. E., and Malcolm, B. R., *Nature* **180**, 1340 (1957).
58. Farrow, G., McIntosh, J., and Ward, I. M., *Makromol. Chem.* **38**, 147 (1960).
59. Farrow, G., and Ward, I. M., *Polymer* **1**, 330 (1960).
60. Ferguson, E. E., *J. Chem. Phys.* **24**, 1115 (1956).
61. Ferington, T. E., and Tobolsky, A. V., *J. Polymer Sci.* **31**, 25 (1958).
62. Field, J. E., Woodford, D. E., and Gehman, S. D., *J. Appl. Phys.* **17**, 386 (1946).
63. Field, N. D., and Schaefgen, J. R., *J. Polymer Sci.* **58**, 533 (1962).
64. Fischmeister, I., and Nilsson, K., *Arkiv Kemi* **16**, 347 (1961).
65. Folt, V. L., Shipman, J. J., and Krimm, S., *J. Polymer Sci.* **61**, S 17 (1962).
66. Fraser, R. D. B., *J. Chem. Phys.* **24**, 89 (1956).
67. Fraser, R. D. B., *in* "Analytical Methods of Protein Chemistry" (P. Alexander and R. J. Block, eds.), Vol. 2, p. 285. Pergamon Press, New York, 1960.
68. Fraser, R. D. B., and MacRae, T. P., *J. Chem. Phys.* **29**, 1024 (1958).
69. Gentilhomme, C., Piguet, A., Rosset, J., and Eyraud, C., *Bull. Soc. Chim. France* p. 901 (1960).
70. Glatt, L., and Ellis, J. W., *J. Chem. Phys.* **15**, 884 (1947).
71. Glatt, L., and Ellis, J. W., *J. Chem. Phys.* **16**, 551 (1948).
72. Glatt, L., and Ellis, J. W., *J. Chem. Phys.* **19**, 449 (1951).
73. Glatt, L., Webber, D. S., Seaman, C., and Ellis, J. W., *J. Chem. Phys.* **18**, 413 (1950).
74. Golub, M. A., and Shipman, J. J., *Spectrochim. Acta* **16**, 1165 (1960).
75. Gotlib, Yu. Ya., *High Mol. Weight Compds. (U.S.S.R.) (English Transl.)* **I**, 474 (1959). [Original in Russian: *Vysokomolekulyarmye Soedinenuya*, **1**, 474 (1959).]
76. Gotlib, Yu. Ya., *Opt. Spectry. (U.S.S.R.) (English Transl.)* **9**, 166 (1960).
77. Gotlib, Yu. Ya., and Kudinskaya, L. V., *Opt. Spectry. (U.S.S.R.) (English Transl.)* **10**, 168 (1961).
78. Gramberg, G., *Angew. Chem.* **73**, 117 (1961).
79. Gramberg, G., *Kolloid-Z.* **175**, 119 (1961).
80. Grime, D., and Ward, I. M., *Trans. Faraday Soc.* **54**, 959 (1958).
81. Grisenthwaite, R. J., and Hunter, R. F., *Chem. & Ind. (London)* p. 719 (1958).
82. Grisenthwaite, R. J., Hunter, R. F., and Krimm, S., *Chem. & Ind. (London)* p. 433 (1959).
83. Haas, H. C., *J. Polymer Sci.* **26**, 391 (1957).
84. Haas, H. C., Emerson, E. S., and Schuler, N. W., *J. Polymer Sci.* **22**, 291 (1956).
85. Hanle, W., Kleinpoppen, H., and Scharmann, A., *Z. Naturforsch.* **13a**, 64 (1958).
86. Hart, E. J., and Meyer, A. W., *J. Am. Chem. Soc.* **71**, 1980 (1949).

87. Harvey, M. C., and Ketley, A. D., *J. Appl. Polymer Sci.* **5**, 247 (1961).

88. Heinen, W., *J. Polymer Sci.* **38**, 545 (1959).

89. Higgins, H. G., *Australian J. Chem.* **10**, 496 (1957).

90. Higgins, H. G., Stewart, C. M., and Harrington, K. J., *J. Polymer. Sci.* **51**, 59 (1961).

91. Higgs, P. W., *J. Chem. Phys.* **23**, 1450 (1955).

92. Hodkins, J. E., *J. Org. Chem.* **23**, 1369 (1958).

93. Holmes, D. R., Miller, R. G., Palmer, R. P., and Bunn, C. W., *Nature* **171**, 1104 (1953).

94. Horne, S. E., Kiehl, J. P., Shipman, J. J., Folt, V. L., Gibbs, C. F., and co-workers, *Ind. Eng. Chem.* **48**, 784 (1956).

95. Huggins, M. L., *J. Chem. Phys.* **13**, 37 (1945).

96. Hummel, D., *Rubber Chem. Technol.* **32**, 854 (1959).

97. Iimura, K., and Takeda, M., *J. Polymer Sci.* **51**, S51 (1961).

98. Immergut, E. H., Kollmanan, G., and Malatesta, A., *J. Polymer Sci.* **51**, S57 (1961).

99. Imoto, S., Ukida, J., and Kominami, T., *Chem. High Polymers (Tokyo)* **14**, 101 (1957).

100. Iozyreva, M. S., *Opt. Spectry. (U.S.S.R.)* (*English Transl.*) **6**, 303 (1959).

101. Ishida, S., *Bull. Chem. Soc. Japan* **33**, 924 (1960).

102. Iwasaki, M., Aoki, M., and Okuhara, K., *J. Polymer. Sci.* **26**, 116 (1957).

103. Jahn, A. S., and Susi, H., *J. Phys. Chem.* **64**, 953 (1960).

104. Jenness, J. R., *J. Opt. Soc. Am.* **50**, 738 (1960).

105. Jones, A. V., Ph.D. Thesis, Cambridge University, 1949.

106. Jones, D. W., *J. Polymer Sci.* **32**, 371 (1958).

107. Jones, R. N., McKay, A. F., and Sinclair, R. G., *J. Am. Chem. Soc.* **74**, 2575 (1952).

108. Kaiser, R., *Kolloid-Z.* **148**, 168 (1956).

109. Kaiser, R., *Kolloid-Z.* **149**, 84 (1956).

110. Kawai, W., and Tsutsumi, S., *Chem. High Polymers (Tokyo)* **18**, 103 (1961).

111. Kawano, M., *J. Chem. Soc. Japan, Pure Chem. Sect.* **82**, 161 (1961).

112. Kawano, M., *J. Chem. Soc. Japan, Pure Chem. Sect.* **82**, 427 (1961).

113. Kawasaki, A., Furukawa, J., Tsuruta, T., Saegsa, T., Kakogawa, G., and Sakata, R., *Polymer* **1**, 315 (1960).

114. Kawasaki, A., Shiotani, S., Furukawa, J., and Tsuruta, T., *Bull. Chem. Soc. Japan* **32**, 1149 (1959).

115. Keller, A., *J. Polymer Sci.* **15**, 31 (1955).

116. Keller, A., and Sandeman, I., *J. Polymer Sci.* **13**, 511 (1954).

117. Keller, A., and Sandeman, I., *J. Polymer Sci.* **15**, 133 (1955).

118. Kellner, L., *Proc. Phys. Soc. (London)* **A64**, 521 (1951).

119. Kern, R. J., *Nature* **187**, 410 (1960).

120. King, W., Hainer, R. M., and McMahon, H. O., *J. Appl. Phys.* **20**, 559 (1949).

121. Kinoshita, Y., *Makromol. Chem.* **33**, 1 (1959).

122. Kislovskii, L. D., *Opt. Spectry. (U.S.S.R.)* (*English Transl.*) **6**, 529 (1959).

123. Kobayashi, M., Nagai, K., and Nagai, E., *Bull. Chem. Soc. Japan* **33**, 1421 (1960).

124. Kriegsmann, H., *Z. Electrochem.* **65**, 336 (1961).

125. Krimm, S., *J. Chem. Phys.* **22**, 567 (1954).

126. Krimm, S., *S.P.E.* (*Soc. Plastics Engrs.*) *J.* **15**, 797 (1959).

127. Krimm, S., *Fortschr. Hochpolymer. Forsch.* **2**, 51 (1960).

128. Krimm, S., *J. Chem. Phys.* **32**, 1780 (1960).

129. Krimm, S., Berens, A. R., Folt, V. L., and Shipman, J. J., *Chem. & Ind. (London)* p. 1512 (1958).

130. Krimm, S., Berens, A. R., Folt, V. L., and Shipman, J. J., *Chem. & Ind. (London)* p. 433 (1959).

131. Krimm, S., and Liang, C. Y., *J. Polymer Sci.* **22**, 95 (1956).

132. Krimm, S., and Liang, C. Y., *J. Appl. Phys.* **29**, 1407 (1958).

133. Krimm, S., Liang, C. Y., and Sutherland, G. B.B. M., *J. Chem. Phys.* **25**, 549 (1956).
134. Krimm, S., Liang, C. Y., and Sutherland, G. B.B. M., *J. Polymer. Sci.* **22**, 227 (1956).
135. Kuroda, Y., and Kubo, M., *J. Polymer Sci.* **26**, 323 (1957).
136. Kuroda, Y., and Kubo, M., *J. Polymer Sci.* **36**, 453 (1959).
137. Larose, P., *Can. J. Chem.* **35**, 1239 (1957).
138. Lenormant, H., Baudras, A., and Blout, E. R., *J. Am. Chem. Soc.* **80**, 6191 (1958).
139. Liang, C. Y., and Krimm, S., *J. Chem. Phys.* **25**, 563 (1956).
140. Liang, C. Y., and Krimm, S., *J. Chem. Phys.* **27**, 327 (1957).
141. Liang, C. Y., and Krimm, S., *J. Chem. Phys.* **27**, 1437 (1957).
142. Liang, C. Y., and Krimm, S., *J. Polymer Sci.* **27**, 241 (1958).
143. Liang, C. Y., and Krimm, S., *J. Polymer Sci.* **31**, 513 (1958).
144. Liang, C. Y., and Krimm, S., *J. Mol. Spectry.* **3**, 554 (1959).
145. Liang, C. Y., and Lytton, M. R., *J. Polymer Sci.* **61**, S45 (1962).
146. Liang, C. Y., Lytton, M. R., and Boone, C. J., *J. Polymer Sci.* **44**, 549 (1960).
147. Liang, C. Y., Lytton, M. R., and Boone, C. J., *J. Polymer Sci.* **54**, 523 (1961).
148. Liang, C. Y., and Marchessault, R. H., *J. Polymer Sci.* **35**, 529 (1959).
149. Liang, C. Y., and Marchessault, R. H., *J. Polymer Sci.* **37**, 385 (1959).
150. Liang, C. Y., and Marchessault, R. H., *J. Polymer Sci.* **39**, 269 (1959).
151. Liang, C. Y., and Marchessault, R. H., *J. Polymer Sci.* **43**, 85 (1960).
152. Liang, C. Y., and Pearson, F. G., *J. Polymer Sci.* **35**, 303 (1959).
153. Liang, C. Y., and Pearson, F. G., *J. Mol. Spectry.* **5**, 290 (1960).
154. Liang, C. Y., Pearson, F. G., and Marchessault, R. H., *Spectrochim. Acta* **17**, 568 (1961).
155. Liang, C. Y., and Watt, W. R., *J. Polymer Sci.* **51**, S14 (1961).
156. Lin, T. P., and Koenig, J. L., *J. Mol. Spectry.* **9**, 228 (1962).
157. Luongo, J. P., *J. Appl. Polymer Sci.* **3**, 302 (1960).
158. Mann, J., and Marrinan, H. J., *J. Polymer Sci.* **32**, 357 (1958).
159. Marchessault, R. H., and Liang, C. Y., *J. Polymer Sci.* **43**, 71 (1960).
160. Marchessault, R. H., and Liang, C. Y., *J. Polymer Sci.* **59**, 375 (1962).
161. Mark, H. F., Atlas, S. M., and Ogata, N., *J. Polymer Sci.* **61**, S49 (1962).
162. Marrinan, H. J., *J. Polymer Sci.* **39**, 461 (1959).
163. Marrinan, H. J., and Mann, J., *J. Polymer Sci.* **21**, 301 (1956).
164. Martin, J. M., Johnston, R. W. B., and O'Neal, M. J., *Spectrochim. Acta* **12**, 12 (1958).
165. Matsumoto, M., and Imai, K., *Chem. High Polymers (Tokyo)* **15**, 160 (1958).
166. Matsuo, H., *Bull. Chem. Soc. Japan* **30**, 593 (1957).
167. Mazzanti, G., Natta, G., Longi, P., and Bernardini, F., *Angew. Chem.* **70**, 601 (1958).
168. McDonald, M. P., and Ward, I. M., *Chem. & Ind. (London)* p. 631 (1961).
169. McMurry, H. L., and Thornton, V., *Anal. Chem.* **24**, 318 (1952).
170. Meiklejohn, R. A., Meyer, R. J., Aronovic, S. M., Schuette, H. A., and Meloch, V. M., *Anal. Chem.* **29**, 329 (1957).
171. Miller, R. G. J., and Willis, H. A., *J. Polymer Sci.* **19**, 485 (1956).
172. Miyake, A., *J. Chem. Soc. Japan, Ind. Chem. Sect.* **62**, 1449 (1959).
173. Miyake, A., *J. Polymer Sci.* **38**, 479 (1959).
174. Miyake, A., *J. Polymer Sci.* **38**, 497 (1959).
175. Miyake, A., *Bull. Chem. Soc. Japan* **33**, 992 (1960).
176. Miyake, A., *J. Phys. Chem.* **64**, 510 (1960).
177. Miyake, A., *J. Polymer Sci.* **44**, 223 (1960).
178. Miyake, A., *Sci. Rept. Toho Rayon Co., Ltd.* **15**, 38 (1960).
179. Miyake, A., *Sci. Rept. Toho Rayon Co., Ltd.* **15**, 68 (1960).
180. Miyake, A., *Sci. Rept. Toho Rayon Co. Ltd.*, **15**, 229 (1960).
181. Miyake, A., *J. Chem. Soc. Japan, Ind. Chem. Sect.* **64**, 710 (1961).

182. Miyake, A., *Sci. Rept. Toho Rayon Co., Ltd.* **16**, 7 (1961).
183. Miyake, A., *Sci. Rept. Toho Rayon Co., Ltd.* **16**, 14 (1961).
184. Miyake, A., *Sci. Rept. Toho Rayon Co., Ltd.* **16**, 91 (1961).
185. Miyazawa, T., *J. Chem. Phys.* **32**, 1647 (1960).
186. Miyazawa, T., *Spectrochim. Acta* **16**, 1231 (1960).
187. Miyazawa, T., *Spectrochim. Acta* **16**, 1233 (1960).
188. Miyazawa, T., *J. Chem. Phys.* **35**, 693 (1961).
189. Miyazawa, T., and Blout, E. R., *J. Am. Chem. Soc.* **83**, 712 (1961).
190. Morero, D., Mantica, E., Ciampelli, F., and Sianesi, D., *Nuovo Cimento* **15**, Suppl. 122 (1960).
191. Morero, D., Mantica, E., and Porri, L., *Nuovo Cimento* **15**, Suppl., 136 (1960).
192. Morero, D., Santambrogio, A., Porri, L., and Ciampelli, F., *Chim. Ind. (Milan)* **41**, 758 (1959).
193. Moynihan, R. E., *J. Am. Chem. Soc.* **81**, 1045 (1959).
194. Nagai, E., Kuribayashi, S., Shiraki, M., and Ukita, M., *J. Polymer Sci.* **35**, 295 (1959).
195. Nagai, E., and Sagane, N., *Chem. High Polymers (Tokyo)* **12**, 195 (1955).
196. Narita, S., Ichinohe, S., and Enomoto, S., *J. Polymer Sci.* **36**, 389 (1959).
197. Narita, S., Ichinohe, S. and Enomoto, S., *J. Polymer Sci.* **37**, 251 (1959).
198. Narita, S., Ichinohe, S., and Enomoto, S., *J. Polymer Sci.* **37**, 263 (1959).
199. Narita, S., Ichinohe, S., and Enomoto, S., *J. Polymer Sci.* **37**, 273 (1959).
200. Natta, G., *Ind. Plastiques Mod. (Paris)* **10**, 40 (1958).
201. Natta, G., Corradini, P., Sianesi, D., and Morero, D., *J. Polymer Sci.* **51**, 527 (1961).
202. Natta, G., Dall'Asta, G., Mazzanti, G., Giannini, U., and Cesca, S., *Angew. Chem.* **71**, 205 (1959).
203. Natta, G., Pasquon, I., Corradini, P., Peraldo, M., Pergoraro, M., and Zambelli, A., *Atti Accad. Nazl. Lincei, Rend., Classe Sci. Fis., Mat. Nat.* **28**, 539 (1960).
204. Natta, G., Porri, L., Corradini, P., Zanini, G., and Ciampelli, F., *Atti Accad. Nazl. Lincei, Rend., Classe Sci. Fis., Mat. Nat.* **29**, 257 (1960).
205. Natta, G., Porri, L., Corradini, P., Zanini, G., and Ciampelli, F., *J. Polymer Sci.* **51** 463 (1961).
206. Natta, G., Sianesi, D., Morero, D., Bassi, I. W., and Caporiccio, G., *Atti Accad. Nazl. Lincei, Rend., Classe Sci. Fis., Mat. Nat.* **28**, 551 (1960).
207. Neuilly, M., *Compt. Rend.* **238**, 65 (1954).
208. Nichols, J. B., *J. Appl. Phys.* **25**, 840 (1954).
209. Nielsen, J. R., and Holland, R. F., *J. Mol. Spectry.* **4**, 488 (1960).
210. Nielsen, J. R., and Holland, R. F., *J. Mol. Spectry.* **6**, 394 (1961).
211. Nielsen, J. R., and Woollett, A. H., *J. Chem. Phys.* **26**, 1391 (1957).
212. Nikitin, V. N., and Volchek, B. Z., *Soviet Phys.—Tech. Phys. (English Transl.)* **2**, 1499 (1957).
213. Novak, A., and Whalley, E., *Can. J. Chem.* **37**, 1710 (1959).
214. Novak, A., and Whalley, E., *Can. J. Chem.* **37**, 1718 (1959).
215. Novak, A., and Whalley, E., *Can. J. Chem.* **37**, 1722 (1959).
216. Novak, A., and Whalley, E., *Trans. Faraday Soc.* **55**, 1484 (1959).
217. Novak, A., and Whalley, E., *Trans. Faraday Soc.* **55**, 1490 (1959).
218. Novak, I. I., *Bull. Acad. Sci. U.S.S.R., Phys. Ser. (English Transl.)* **22**, 1103 (1958).
219. O'Connor, R. T., DuPré, E. F., and McCall, E. R., *Anal. Chem.* **29**, 998 (1957).
220. O'Connor, R. T., DuPré, E. F., and Mitcham, D., *Textile Res. J.* **28**, 382 (1958).
221. O'Connor, R. T., McCall, E. R., and Mitcham, D., *Am. Dyestuff Reptr.* **49**, 35 (1960).
222. Ohshika, T., *J. Chem. Soc. Japan, Ind. Chem. Sect.* **63**, 1256 (1960).
223. Onishi, T., and Krimm, S., *J. Appl. Phys.* **32**, 2320 (1961).

224. Palm, A., *J. Phys. Chem.* **55**, 1320 (1951).
225. Patterson, D., and Ward, I. M. *Trans. Faraday Soc.* **53** 291 (1957).
226. Pearson, F. G., Marchessault, R. H. and Liang C. Y., *J. Polymer Sci.* **43**, 101 (1960).
227. Peraldo, M., *Gazz. Chim. Ital.* **89**, 798 (1959).
228. Peraldo, M., and Farina, M., *Chim. Ind. (Milan)* **42**, 1349 (1960).
229. Perron, R., and Perichon, J., *Compt. Rend.* **252**, 3224 (1961).
230. Pimentel, G. C., and Klemperer, W. A., *J. Chem. Phys.* **23**, 376 (1955).
231. Pohl, H. A., Bacskai, R., and Purcell, W. P., *J. Phys. Chem.* **64**, 1701 (1960).
232. Pokrovsky, E. I., *Zh. Fiz. Khim.* **32**, 1410 (1958).
233. Pokrovsky, E. I., and Kotova, I. P., *Soviet Phys.—Tech. Phys. (English Transl.)* **1**. 1417 (1957).
234. Pokrovsky, E. I., and Volkenshtein, M. V., *Proc. Acad. Sci. U.S.S.R., Phys. Chem. Sect. (English Transl.)* **115**, 517 (1957).
235. Primas, H., and Günthard, H. H., *Helv. Chim. Acta* **36**, 1659 (1953).
236. Primas, H., and Günthard, H. H., *Helv. Chim. Acta* **36**, 1791 (1953).
237. Quynn, R. G., Ph.D. Thesis, Princeton University, 1957.
238. Quynn, R. G., Riley. J. L., Young, D. A., and Noether, H. D., *J. Appl. Polymer Sci.* **2**, 166 (1959).
239. Reding, F. P., and Brown, A., *J. Appl. Phys.* **25**, 848 (1954).
240. Reding, F. P., and Lovell, C. M., *J. Polymer Sci.* **21**, 157 (1956).
241. Richardson, W. S., and Sacher, A., *J. Polymer Sci.* **10**, 353 (1953).
242. Robinson, T. S., and Price, W. C., *Proc. Phys. Soc. (London)* **B66**, 969 (1953).
243. Rugg, F. M., Smith, J. J., and Atkinson, J. V., *J. Polymer Sci.* **9**, 579 (1952).
244. Salomon, G., Schooneveldt-van der Kloes, C. J., and Zwiers, J. H. L., *Rec. Trav. Chim.* **79**, 313 (1960).
245. Salomon, G., Van der Schee, A. C., Ketelaar, J. A. A., and Van Eyk, B. J., *Rubber Chem. Technol.* **25**, 107 (1952).
246. Schaefgen, J. R., and Zbinden, R., *J. Polymer Sci.* (to be published).
247. Schnell, G., *Ergeb. Exakt. Naturw.* **31**, 270 (1959).
248. Schnell, H., *Plastics Inst. (London), Trans. J.* **28**, 143 (1960).
249. Schulz, R. C., Cherdron, H., and Kern, W., *Makromol. Chem.* **29**, 190 (1959).
250. Schurz, J., Bayzer, H., and Stübchen, H., *Makromol. Chem.* **23**, 152 (1957).
251. Seidel, B., *Z. Elektrochem.* **62**, 214 (1958).
252. Sheppard, N., *Advan. Spectry.* **1**, 288 (1959).
253. Shimanouchi, T., and Mizushima, S., *Sci. Papers Inst. Phys. Chem. Res. (Tokyo)* **40**, 467 (1943).
254. Shimanouchi, T., and Tasumi, M., *Bull. Chem. Soc. Japan* **34**, 359 (1961).
255. Shimanouchi, T., and Tasumi, M., *J. Chem. Phys.* **34**, 687 (1961).
256. Shimanouchi, T., and Tasumi, M., *Spectrochim. Acta* **17**, 755 (1961).
257. Shimanouchi, T., Tsuchiya, S., and Mizushima, S., *J. Chem. Phys.* **30**, 1365 (1959).
258. Shipman, J. J., and Golub. M. A., *J. Polymer Sci.* **58**, 1063 (1962).
259. Simon, A., Mücklich, M., Kunath, D., and Heintz, G., *J. Polymer Sci.* **30**, 201 (1958).
260. Singer, R., and Weiler, J., *Helv. Chim. Acta* **15**, 649 (1932).
261. Skoda, W., and Schurz, J., *Makromol. Chem.* **29**, 156 (1959).
262. Slichter, W. P., and Mandell, E. R., *J. Appl. Phys.* **29**, 1438 (1958).
263. Snyder, R. G., *J. Chem. Phys.* **27**, 969 (1957).
264. Snyder, R. G., *J. Mol. Spectry.* **4**, 411 (1960).
265. Snyder, R. G., *J. Mol. Spectry.* **7**, 116 (1961).
266. Spach, G., and Horn, P., *Compt. Rend.* **248**, 399 (1959).
267. Starkweather, H. W., and Moynihan, R. E., *J. Polymer Sci.* **22**, 363 (1956).
268. Stavely, F. W. and co-workers, *Ind. Eng. Chem.* **48**, 778 (1956).

269. Stein, R. S., *J. Chem. Phys.* **23**, 734 (1955).
270. Stein, R. S., *J. Polymer Sci.* **31**, 327 (1958).
271. Stein, R. S., *J. Polymer Sci.* **31**, 335 (1958).
272. Stein, R. S., *J. Polymer Sci.* **34**, 709 (1959).
273. Stein, R. S., and Norris, F. H., *J. Polymer Sci.* **21**, 381 (1956).
274. Stein, R. S., and Sutherland, G. B. B. M., *J. Chem. Phys.* **21**, 370 (1953).
275. Stein, R. S., and Sutherland, G. B. B. M., *J. Chem. Phys.* **22**, 1993 (1954).
276. Stepanov, B. I., Zhbankov, R. G., and Rozenberg, A. Ya., *Zh. Fiz. Khim.* **33**, 1907 (1959).
277. Stewart, J. E., and Hellman, M., *J. Res. Natl. Bur. Std.* **60**, 125 (1958).
278. Strasheim, A., and Buijs, K., *Spectrochim. Acta* **16**, 1010 (1960).
279. Stuart, H. A., and Zachmann, H. G., *Angew. Chem.* **73**, 245 (1961).
280. Susi, H., *Anal. Chem.* **31**, 910 (1959).
281. Susi, H., *J. Am. Chem. Soc.* **81**, 1535 (1959).
282. Susi, H., and Smith, A. M., *J. Am. Oil Chemists' Soc.* **37**, 431 (1960).
283. Sutherland, G. B. B. M., and Jones, A. V., *Discussions Faraday Soc.* **9**, 281 (1950).
284. Sutherland, G. B. B. M., and Sheppard, N., *Nature* **159**, 739 (1947).
285. Sydow, E. von, *Acta Chem. Scand.* **9**, 1119 (1955).
286. Szigeti, B., *Proc. Roy. Soc.* **A264**, 198 (1961).
287. Szigeti, B., *Proc. Roy. Soc.* **A264**, 212 (1961).
288. Tadokoro, H., *Bull. Chem. Soc. Japan* **32**, 1252 (1959).
289. Tadokoro, H., *Bull. Chem. Soc. Japan* **32**, 1334 (1959).
290. Tadokoro, H. Kitazawa, T., Nozakura, S., and Murahashi, S., *Chem. High Polymers (Tokyo)* **17**, 231 (1960).
291. Tadokoro, H., Kitazawa, T., Nozakura, S., and Murahashi, S., *Bull. Chem. Soc. Japan* **34**, 1209 (1961).
292. Tadokoro, H., Kobayashi, A., Kawaguchi, Y., Sobajima, S., Murahashi, S., and Matsui, Y., *J. Chem. Phys.* **35**, 369 (1961).
293. Tadokoro, H., Kobayashi, M., Murahashi, S., Mitsuishi, A., and Yoshinaga, H., *Bull. Chem. Soc. Japan* **35**, 1429 (1962).
294. Tadokoro, H., Kozai, K., Seki, S., and Nitta, I., *J. Polymer Sci.* **26**, 379 (1957).
295. Tadokoro, H., Nagai, H., Seki, S., and Nitta, I., *Bull. Chem. Soc. Japan* **34**, 1504 (1961).
296. Tadokoro, H., Nishiyama, N., Nozakura, S., and Murahashi, S., *J. Polymer Sci.* **36**, 553 (1959).
297. Tadokoro, H., Nishiyama, Y., Nozakura, S., and Murahashi, S., *Bull. Chem. Soc. Japan* **34**, 381 (1961).
298. Tadokoro, H., Nozakura, S., Kitazawa, T., Yasuhara, Y., and Murahashi, S., *Bull. Chem. Soc. Japan* **32**, 313 (1959).
299. Tadokoro, H., Seki, S., and Nitta, I., *J. Chem. Phys.* **23**, 1351 (1955).
300. Tadokoro, H., Seki, S., and Nitta, I., *J. Polymer Sci* **22**, 563 (1956).
301. Tadokoro, H., Seki, S., Nitta, I., and Yamadera, R., *J. Polymer Sci.* **28**, 244 (1958).
302. Tadokoro, H., Tatsuka, K., Murahashi, S., *J. Polymer Sci.* **59**, 413 (1962).
303. Takeda, M., Iimura, K., Yamada, A., and Imamura, Y., *Bull. Chem. Soc. Japan* **32**, 1150 (1959).
304. Takeda, M., Iimura, K., Yamada, A., and Imamura, Y., *Bull. Chem. Soc. Japan* **33**, 1219 (1960).
305. Tasumi, M., and Shimanouchi, T., *Spectrochim. Acta* **17**, 731 (1961).
306. Thompson, H. W., and Torkington, P. *Proc. Roy. Soc.* **A184** 3 (1945).
307. Thompson, H. W., and Torkington, P., *Trans. Faraday Soc.* **41**, 246 (1945).
308. Tobin, M. C., *J. Chem. Phys.* **23**, 891 (1955).
309. Tobin, M. C., *J. Phys. Chem.* **61**, 1392 (1957).

310. Tobin, M. C., *J. Phys. Chem.* **64**, 216 (1960).
311. Tobin, M. C., and Carrano, M. J., *J. Chem. Phys.* **25**, 1044 (1956).
312. Tobin, M. C., and Carrano, M. J., *J. Polymer Sci.* **24**, 93 (1957).
313. Tschamler, H., *J. Chem. Phys.* **22**, 1845 (1954).
314. Tsuboi, M., *J. Polymer Sci.* **25**, 159 (1957).
315. Tsuboi, M., *J. Polymer Sci.* **59**, 139 (1962).
316. Ueberreiter, K., and Krull, W., *Angew. Chem.* **69**, 557 (1957).
317. Volchek, B. Z., and Nikitin, V. N., *Soviet Phys.—Tech. Phys.* (*English Transl.*) **2**, 1705 (1957).
318. Volchek, B. Z., and Nikitin, V. N., *Soviet Phys.—Tech. Phys.* (*English Transl.*) **3**, 1617 (1958).
319. Ward, I. M., *Chem. & Ind.* (*London*) p. 905 (1956).
320. Ward, I. M., *Chem. & Ind.* (*London*) p. 1102 (1957).
321. Ward, I. M., *Trans. Faraday Soc.* **53**, 1406 (1957).
322. Wentink, T., and Martin, C. E., *J. Opt. Soc. Am.* **50**, 741 (1960).
323. Wentink, T., Willwerth, L. J., and Phaneuf, J. P., *J. Polymer Sci.* **55**, 551 (1961).
324. Wenzel, F., Schiedt, U., and Breusch, F. L., *Z. Naturforsch.* **12b** 71 (1957).
325. White, H. F., and Lovell, C. M., *J. Polymer Sci.* **41**, 369 (1959).
326. Wood, D. L., and Luongo, J. P., *Mod. Plastics* **38**, 132 (1961).
327. Woollett, A. H., Ph.D. Thesis, University of Oklahoma, 1956.
328. Yamadera, R., *J. Polymer Sci.* **50**, S4 (1961).
329. Yen, T. F., *J. Polymer Sci.* **35**, 533 (1959).
330. Zachmann, H. G., and Stuart, H. A., *Makromol. Chem.* **44/46**, 622 (1961).
331. Zahn, H., and Seidel, B., *Makromol. Chem.* **29**, 70 (1959).
332. Zbinden, R., *J. Mol. Spectry.* **3**, 654 (1959).

Author Index

Shipman, J. J., 25(40), *31*, 235(65), 237 (129, 130), 240(74, 94, 258), *245, 246, 249*

Shiraki, M., 236(194), *248*

Shivers, J. C., 3(17), *31*

Shrestha, C. B., 75(26), *97*

Shurcliff, W. A., 168(9), *232*

Sianesi, D., 228(42), *232*, 235(190, 201), 236(206), *248*

Siegler, E. H., 169, 170(19), 171(19), *232*

Simon, A., 235(259), 236(259), 237(259), *249*

Simpson, D. M., 131, 133(7, 8), 148, 149 (7, 36), 151(8), 155, *164*, 235(29, 30), *244*

Sinclair, R. C., 5(74), *32*

Sinclair, R. G., 148, 155(21), *164*, 235 (107), *246*

Singer, R., 235(260), *249*

Skoda, W., 238(261), *249*

Slichter, W. P., 235(262), *249*

Smith, A. E., 153(37), *164*

Smith, A. M., 155(42), *164*, 235(282), *250*

Smith, J. J., 22(72), *32*, 234(243), *249*

Snyder, R. G., 74(30), 87(31), *97*, 149(39), 150(38), 151(38, 39), 153(38), 155, *164*, 235(263, 264, 265), *249*

Sobajima, S., 124(23), *128*, 240(292), *250*

Sorenson, W. R., 1(75), *32*

Spach, G., 243(266), *249*

Speiser, A., 33(32), *97*

Starkweather, H. W., 16, 17(76), *32*, 242 (267), *249*

Statton, W. O., 10(10), 11(10), *31*, 119 (20, 21), *128*, 237(20), *244*

Stavely, F. W., 25(77), *32*, 240(268), *249*

Stein, R. S., 180(53), *233*, 234(270, 271, 272, 273, 275), 235(269, 274, 275), *250*

Stepanov, B. I., 243(276), *250*

Stewart, C. M., 243(90), *246*

Stewart, J. E., 243(277), *250*

Strasheim, A., 243(278), *250*

Stross, F. H., 21(11), *31*

Strutters, G. W., 155(11), *164*

Stuart, H. A., 1(78), *32*, 234(279, 330), 241(330), *250, 251*

Stübchen, H., 237(250), 238(250), *249*

Susi, H., 155(20, 40, 41, 42), 156, *164*, 235 (103, 280, 281, 282), *246, 250*

Sutherland, G. B. B. M., 4(51), *32*, 81(18), 85(33), 86, 87(18), 89(18), *97*, 104(9), 110, 122(9), *128*, 149, *164*, 234 (133, 275, 284), 235(274, 275), 236 (134), 240(283), 243(14), *244, 247, 250*

Sweeting, O. J., 234(2), *244*

Swenson, C. A., 4(64, 79, 80), *32*

Szigeti, B., 4(81), *32*, 235(286, 287), *250*

T

Tadokoro, H., 48(34), *97*, 124(22, 23), *128*, 228(55), *233*, 235(290, 291, 296, 297, 298), 236(288, 289, 291, 294, 295, 298, 299, 300, 301), 240(292, 293), 241(302), *250*

Takeda, M., 235(303, 304), 237(97), *246, 250*

Tanner, K. N., 243(14), *244*

Tasumi, M., 237(254, 255, 256, 305), *249, 250*

Tatsuka, K., 241(302), *250*

Temple, R. B., 167(22), 169(23), *232*, 234 (5, 56), 236(5, 56), 237(5, 56), 242(5, 56), 243(4), *244, 245*

Tessmar, K., 28(3), *30*, 238(12), *244*

Theimer, O. H., 162, *164*

Thomas, L. H., 131, *165*

Thompson, H. W., 24(82), 25(82), *32*, 169(4), *232*, 234(306), 235(306, 307), 236(307), 237(306), 238(307), 240(306), *250*

Thornton, V., 153(24), *164*, 235(169), *247*

Till, P. H., 2(83), *32*

Tilley, G. P., 234(2), *244*

Tink, R. R., 179, *233*

Tobin, M. C., 39(35), 46(35), 50, 80, 81(35), 86(35), *97*, 185(57), 212, *233*, 234(308, 309, 311, 312), 235(310, 311), 241(309), 242(311), 243(309), *250, 251*

Tobolsky, A. V., 240(61), *245*

Torkington, P., 24(82), 25(82), *32*, 234 (306), 235(306, 307), 236(307), 237(306), 238(307), 240(306), *250*

Tschamler, H., 131, 133(48), 149(48), *165*, 235(313), *251*

Tsuboi, M., 243(9, 314, 315), *244, 251*

Tsuchiya, S., 237(257), *249*

Tsuruta, T., 237(114), 241(113), *246*

Tsutsumi, S., 238(110), *246*

Subject Index

A

Abelian group, 34, 43
Analysis, quantitative, 26
Analytical applications, 18 ff.
Antiparallel dipoles, 135 ff.
Atactic polymers, 1
Axial orientation, 190, 195 ff.

B

Band series, 141 ff.
Beam condensing systems, 169 ff.
Beam convergence, 177 ff.
Birefringence, 212
Brillouin zone, 68

C

Carbon arc, 169
Cellulose, 243
 repeat unit, 1
Chain of point masses, 98 ff.
Character of a representation, 38 ff.
Character table
 one-dimensional translation group, 44
 point group, 41
$C_{23}H_{48}$ spectrum, 150
$C_{24}H_{50}$ spectrum, 150
CH_2 vibrations, 148
 antisymmetric stretching, 148, 158
 bending, 148, 155
 rocking, 148, 149 ff.
 symmetric stretching, 148, 158
 twisting, 145, 148
 wagging, 148, 155 ff.
Class, 34, 40
Combination bands, 79
Configurational isomers, 27 ff.
Conjugate element, 34
Coordinate transformation, 181 ff.
Coset, 35
Coupled oscillators, 5, 129 ff.
 equation of motion, 132
 form of normal vibrations, 136 ff.
 frequency distribution, 134 ff.

Coupled oscillators—*continued*
 limitations of theory, 146 ff.
 magnitude of splitting, 136
Crystallinity, 2, 149, 166
 effect on spectrum, 11 ff.
Cross-linking, 20

D

Diatomic linear lattice, 105 ff.
 difference bands, 68 ff., 105 ff.
 frequency branches, 106 ff.
 longitudinal vibrations, 105 ff.
 transverse vibrations, 108 ff.
Dichroic coefficient, 179 ff.
Dichroic ratio, 7, 8, 172 ff., 185 ff.
Dichroism, 8 ff., 166 ff.
Difference bands, 68 ff.
Dipole moment change, 142 ff., 180
Distribution function, 8, 182 ff., 198 ff.
Drawn sample, 167, 207 ff.

E

Eigenfunction, 63
Electron diffraction, 194, 206, 209, 212, 221
Ellipsoid, 182 ff.
End group, 1
 analysis, 20 ff.
Energy barrier for rotation, 149, 154

F

Factor group, 35
 of line group, 47 ff.
 of space group, 45 ff.
 vibrations, 6
False radiation, 170 ff., 176
Fatty acids, 235
Fiber, 167
Fibrous samples, 169 ff.
Film, 166, 213 ff.
Fixed boundary model, 99 ff.

261